# Quantum Machine Learning

Concepts and possibilities

**Online at: https://doi.org/10.1088/978-0-7503-4952-9**

# IOP Series in Quantum Technology

**Series Editor**: **Barry Garraway** (School of Mathematical and Physical Sciences, University of Sussex, UK), **Barry Sanders** (Institute for Quantum Science and Technology, University of Calgary, Canada) and **Lincoln Carr** (Quantum Engineering Program, Colorado School of Mines, USA)

## About the Series

The IOP Series in Quantum Technology is dedicated to bringing together the most up to date texts and reference books from across the emerging field of quantum science and its technological applications. Prepared by leading experts, the series is intended for graduate students and researchers either already working in or intending to enter the field. The series seeks (but is not restricted to) publications in the following topics:

- Quantum biology
- Quantum communication
- Quantum computation
- Quantum control
- Quantum cryptography
- Quantum engineering
- Quantum machine learning and intelligence
- Quantum materials
- Quantum metrology
- Quantum optics
- Quantum sensing
- Quantum simulation
- Quantum software, algorithms and code
- Quantum thermodynamics
- Hybrid quantum systems

A list of recent titles published in this series can be found here: https://iopscience.iop.org/bookListInfo/iop-series-in-quantum-technology.

# Quantum Machine Learning

Concepts and possibilities

**Andrea Delgado**
*Physics Division, Oak Ridge National Laboratory, Oak Ridge, TN, USA*

**Kathleen E Hamilton**
*Computational Sciences and Engineering Division, Oak Ridge National Laboratory, Oak Ridge, TN, USA*

**IOP** Publishing, Bristol, UK

ISBN 978-0-7503-4952-9 (ebook)
ISBN 978-0-7503-4950-5 (print)
ISBN 978-0-7503-4953-6 (myPrint)
ISBN 978-0-7503-4951-2 (mobi)

DOI 10.1088/978-0-7503-4952-9

Version: 20251201

IOP ebooks

British Library Cataloguing-in-Publication Data: A catalogue record for this book is available from the British Library.

Published by IOP Publishing, wholly owned by The Institute of Physics, London

IOP Publishing, No.2 The Distillery, Glassfields, Avon Street, Bristol, BS2 0GR, UK

US Office: IOP Publishing, Inc., 190 North Independence Mall West, Suite 601, Philadelphia, PA 19106, USA

# Contents

**Preface** x

**Author biographies** xi

**1 Introduction** **1-1**

1.1 Motivation and scope 1-1
1.2 The promise of quantum machine learning 1-2
1.3 Quantum computing: a refresher 1-3
1.4 Machine learning foundations 1-4
1.5 Types of quantum machine learning 1-5
References 1-7

**2 Quantum information processing** **2-1**

2.1 Quantum computing 2-1
2.2 Dirac notation 2-3
2.3 Bloch sphere 2-3
2.4 The Copenhagen interpretation 2-4
2.5 Quantum gates and circuits 2-5
2.5.1 Quantum gates 2-5
2.5.2 The quantum circuit model 2-6
2.6 The Deutsch–Jozsa algorithm 2-7
2.6.1 Problem statement 2-7
2.6.2 Algorithm overview 2-8
2.6.3 Interpretation of results 2-8
2.6.4 Significance 2-8
2.7 Grover's algorithm 2-8
2.7.1 Problem statement 2-8
2.7.2 Algorithm overview 2-9
2.7.3 Mathematical insight 2-9
2.7.4 Significance 2-9
2.8 Summary 2-9
References 2-10

**3 Information encoding** **3-1**

3.1 Bridging two computational realms 3-2
3.2 Encoding schemes for quantum machine learning 3-4
3.2.1 Amplitude encoding 3-4

3.2.2 Angle/parametric encoding 3-5
3.2.3 Basis encoding 3-5
3.2.4 Hamiltonian encoding 3-5
3.3 Quantum autoencoders 3-7
3.3.1 Circuit construction 3-7
3.3.2 Applications and relevance 3-8
3.4 Quantum data 3-8
3.4.1 Definition and representations 3-9
3.4.2 Sources of quantum data 3-9
3.4.3 Quantum learning from quantum data 3-10
3.4.4 Implications for encoding 3-10
3.5 Practical considerations for quantum encoding 3-11
3.6 Summary 3-12
References 3-13

**4 Quantum computing for inference 4-1**

4.1 Classical foundations of inference 4-2
4.1.1 Supervised learning basics 4-2
4.1.2 Feature maps and kernels in classical machine learning 4-2
4.1.3 Kernel methods 4-3
4.1.4 Theoretical guarantees 4-3
4.2 Quantum feature maps 4-4
4.2.1 Quantum states as features 4-4
4.2.2 Data encodings for quantum feature maps 4-5
4.2.3 Properties of quantum feature maps 4-5
4.3 Quantum kernels and kernel methods 4-6
4.3.1 Quantum kernel definition 4-6
4.3.2 Evaluating quantum kernels 4-7
4.3.3 Training with quantum kernels 4-7
4.3.4 Generalization and expressivity 4-7
4.4 Linear quantum models 4-8
4.4.1 Parameterized quantum circuits for classification 4-8
4.4.2 Training strategies 4-9
4.4.3 Example: variational quantum classifier 4-9
4.5 Performance and benchmarks 4-10
4.5.1 Theoretical speedups and limitations 4-10
4.5.2 Empirical studies and benchmarks 4-10

| | | |
|---|---|---|
| | 4.5.3 Robustness and noise considerations | 4-11 |
| 4.6 | Case studies and applications | 4-11 |
| | 4.6.1 Drug discovery and molecular property prediction | 4-12 |
| | 4.6.2 High energy physics | 4-12 |
| | 4.6.3 Finance | 4-12 |
| 4.7 | Open challenges and future directions | 4-13 |
| | 4.7.1 Scalability and resource constraints | 4-13 |
| | 4.7.2 Design of quantum feature maps | 4-14 |
| | 4.7.3 Noise robustness and error mitigation | 4-14 |
| | 4.7.4 Learning theory and generalization bounds | 4-14 |
| | 4.7.5 Hybrid quantum–classical workflows | 4-15 |
| 4.8 | Summary | 4-15 |
| | References | 4-16 |
| **5** | **Quantum variational optimization** | **5-1** |
| 5.1 | Model description | 5-2 |
| | 5.1.1 Variational quantum eigensolver | 5-2 |
| | 5.1.2 Quantum approximate optimization algorithm | 5-3 |
| 5.2 | Case studies and applications | 5-4 |
| | 5.2.1 Quantum chemistry and materials science | 5-4 |
| | 5.2.2 Combinatorial optimization | 5-4 |
| | 5.2.3 Quantum machine learning | 5-5 |
| | 5.2.4 Quantum control and metrology | 5-6 |
| | 5.2.5 Quantum simulation and dynamics | 5-6 |
| 5.3 | Open challenges and future directions | 5-6 |
| 5.4 | Summary | 5-9 |
| | References | 5-9 |
| **6** | **Variational classifiers and neural networks** | **6-1** |
| 6.1 | Model description | 6-2 |
| | 6.1.1 Classical neural networks | 6-2 |
| | 6.1.2 Parameterized quantum circuits as learning models | 6-3 |
| 6.2 | Backpropagation and gradient estimation in quantum models | 6-4 |
| | 6.2.1 Automatic differentiation on hybrid computational graphs | 6-5 |
| | 6.2.2 Parameter-shift rule | 6-7 |
| | 6.2.3 Generalized parameter-shift rules | 6-8 |

6.2.4 Finite differences and simultaneous perturbation stochastic approximation 6-10
6.2.5 Adjoint differentiation (simulators only) 6-12
6.2.6 Quantum natural gradient and the quantum geometric tensor 6-13
6.2.7 Comparison to the classical natural gradient 6-15
6.2.8 Gradients for mixed states, noisy channels, and open-system dynamics 6-15
6.2.9 Gradient variance, shot noise, and batching 6-17
6.2.10 Loss landscapes and barren plateaus 6-19
6.2.11 Intuition behind barren plateaus 6-19
6.3 Architectural variants: QCNNs and quantum graph-based models 6-20
6.3.1 Quantum convolutional neural networks 6-21
6.3.2 Quantum graph neural networks 6-22
6.4 Open challenges and future directions 6-24
6.4.1 Trainability and optimization 6-24
6.4.2 Resource overhead and compilation 6-24
6.4.3 Generalization and inductive bias 6-25
6.4.4 Benchmarking and expressivity analysis 6-25
6.4.5 Opportunity in co-design 6-25
6.5 Summary 6-25
References 6-26

**7 Quantum generative models 7-1**
7.1 Classical generative models 7-2
7.1.1 Boltzmann machines and energy-based models 7-2
7.1.2 Deep generative models 7-3
7.2 Quantum generative models 7-4
7.2.1 Quantum Boltzmann machines 7-5
7.2.2 Quantum circuit Born machines 7-7
7.2.3 Quantum generative adversarial networks 7-8
7.2.4 Quantum variational autoencoders 7-8
7.3 Expressivity and learning power 7-8
7.3.1 Defining expressivity 7-9
7.3.2 Factors affecting expressivity 7-9
7.3.3 Trade-off with trainability 7-10
7.4 Training quantum generative models 7-10
7.4.1 The optimization loop 7-10
7.4.2 Loss functions 7-11

7.4.3 Gradient estimation 7-11
7.4.4 Training challenges 7-11
7.5 Case studies and applications 7-12
7.5.1 Learning classical data distributions 7-12
7.5.2 Quantum state learning and tomography 7-12
7.5.3 Quantum circuit compilation and data compression 7-13
7.5.4 Anomaly detection 7-13
7.5.5 Physical simulation and sampling 7-13
7.6 Open challenges and future directions 7-14
7.6.1 Challenges 7-14
7.6.2 Evaluation metrics and benchmarks 7-15
7.6.3 Opportunities 7-15
7.7 Summary 7-16
References 7-17

**8 Theory, expressivity, and learning bounds 8-1**

8.1 Definitions and frameworks for expressivity 8-2
8.1.1 Hypothesis classes 8-2
8.1.2 Quantifying expressivity 8-3
8.1.3 Expressivity versus trainability 8-4
8.2 Learning performance: sample complexity and generalization 8-6
8.2.1 PAC and agnostic learning 8-6
8.2.2 Quantum sample complexity 8-7
8.2.3 Generalization bounds for quantum models 8-8
8.3 Positive results from quantum learning 8-9
8.3.1 Query and time complexity speedups 8-9
8.3.2 Quantum kernels and feature maps 8-10
8.3.3 Quantum generative models 8-10
8.4 Limitations, no-go theorems, and dequantization 8-12
8.4.1 Sample complexity lower bounds 8-12
8.4.2 Barren plateaus and optimization barriers 8-13
8.4.3 Dequantization results 8-13
8.4.4 Robustness and noise sensitivity 8-14
8.5 Open problems and future directions 8-14
References 8-15

# Preface

Quantum computing and machine learning are some of today's most active and rapidly expanding fields of research and development. Quantum machine learning has evolved in recent years as a subdiscipline of quantum computing that explores how quantum computers can be used for machine learning tasks—in other words, how quantum computers can learn from data, either quantum or classical. Quantum machine learning is an exciting field because it holds the promise of accelerating classical machine learning algorithms.

This book will showcase the growing number of algorithms and practical applications of quantum machine learning. Starting from the foundational approach that reformulates learning theory in a quantum setting, this book will highlight recent efforts to find quantum algorithms that speed up machine learning regarding computational complexity measures. The book will also cover new machine learning applications tailored for noisy intermediate-scale quantum (NISQ) devices from a near-term or co-design perspective, including hybrid quantum–classical methods. The aim of this book is to help readers develop a clear understanding of the promises and limitations of the current state-of-the-art of quantum machine learning algorithms, and help them define directions for future research in this exciting field. We start by looking at the quantum mechanical principles that enable quantum computing and helping the reader gather some intuition about how to harness these quantum mechanical properties to speed up some computational tasks. Then, we introduce the reader to some of the basic quantum algorithms that are often used to develop more complicated algorithms. Later, we explore the different fields of study in quantum machine learning, which are related to the choice of algorithm, and whether it operates on quantum or classical data. Finally, we describe the state-of-the-art for quantum machine learning algorithms including both supervised and unsupervised learning, discussing their applications and limitations for implementation on near-term devices.

Quantum machine learning is a subject in the making, with endless possibilities for applications in the near and long term. Nonetheless, to find out what quantum machine learning has to offer, its numerous possible avenues first have to be explored by an interdisciplinary community of scientists and quantum computing enthusiasts. We intend this book to be a possible starting point for this journey, as it is intended to introduce some key concepts, ideas, and algorithms that are the result of the first few years of quantum machine learning research. Our aim is to provide a comprehensive literature review and to summarize key topics that appear often in quantum machine learning, to put them into context and make them accessible to a broader audience in order to foster future research and applications.

# Author biographies

## Dr Andrea Delgado

**Dr Andrea Delgado** is a Research Scientist in the Physics Division at Oak Ridge National Laboratory. Her research focuses on quantum computing applications to high-energy physics, with an emphasis on developing novel algorithms for data analysis and simulation in large-scale particle physics experiments, such as those at the Large Hadron Collider. She is a recipient of the prestigious Eugene P Wigner Fellowship, awarded for her pioneering work at the intersection of quantum information science and fundamental physics. Dr Delgado serves on the steering committee of the IBM/CERN Quantum Computing for High Energy Physics Working Group and has played a leadership role in IEEE Quantum Week since its inception, including multiple terms as Technical Program Chair. She is also an Associate Editor for *IEEE Transactions on Quantum Engineering*. Dr Delgado received her PhD in Physics from Texas A&M University, where she specialized in searches for new physics. Her current work explores quantum machine learning, quantum generative modeling, and hybrid quantum–classical algorithms with the aim of advancing discovery in high-energy and nuclear physics.

## Dr Kathleen E Hamilton

**Dr Kathleen E Hamilton** is a Research Scientist in the Quantum Computational Science Group at Oak Ridge National Laboratory. Her research interests cover the development of benchmarks and algorithms for next-generation processors in quantum computing or neuromorphic computing. Her work in quantum machine learning include designing new approaches for natural language processing, reservoir computing, and using machine learning workflows to benchmark near-term quantum devices. She has served on the Program Committee for the International Conference on Neuromorphic Systems (ICONS) from 2019 to 2021, and has co-organized the Quantum Artificial Intelligence Workshop which has been a part of IEEE's Quantum Week since 2020. She received her PhD from the University of California at Riverside.

**IOP** Publishing

Quantum Machine Learning

Concepts and possibilities

**Andrea Delgado and Kathleen E Hamilton**

# Chapter 1

# Introduction

## 1.1 Motivation and scope

The accelerating development of quantum technologies has opened new frontiers in computation, communication, and sensing. Among the most promising direction is **quantum machine learning (QML)**.

QML aims to explore whether and how quantum resources can enhance learning tasks, improve model expressivity, or accelerate training and inference. It combines the principles of quantum mechanics and machine learning to explore new ways of processing, analyzing, and generating data.

This book introduces the reader to the core ideas, algorithms, and challenges in QML, presenting both theoretical foundations and practical applications tailored for current and near-term quantum devices. The text is structured to serve as a **self-contained textbook**, suitable for graduate students or researchers with a working knowledge of quantum computing and a general background in classical machine learning.

**Quantum speedup in learning**

QML is motivated in part by the potential for **computational speedups** over classical algorithms [1]. But what does 'speedup' mean in this context? In general, a quantum algorithm exhibits a *quantum speedup* when it solves a problem with lower resource usage (e.g. time, queries, or samples) than any known classical algorithm [2]. These speedups can take various forms:

- **Polynomial speedup:** Occurs when a quantum algorithm reduces the runtime from $O(n^k)$ to $O(n^m)$ for $m < k$. *Example*: Grover's algorithm [3] for unstructured search achieves a quadratic speedup ($O(N)$ to $O(\sqrt{N})$).
- **Exponential speedup:** Occurs when a quantum algorithm solves a task in time poly($n$) while classical solutions require exp($n$). *Example*: Shor's factoring algorithm [4].

doi:10.1088/978-0-7503-4952-9ch1

- **Super-polynomial speedup in query complexity:** In some learning-theoretic settings, some concept classes can be learned with exponentially fewer queries by quantum learners. For example, certain parity functions can be learned with exponentially fewer queries on a quantum computer [5, 6].

**Applications to learning theory:** Let us consider a few scenarios where quantum speedup is either proven or conjectured in the context of learning:

- *PAC learning:* In the probably approximately correct (PAC) framework, quantum learners can improve sample complexity or access richer hypothesis classes when provided with quantum examples [7, 8].
- *Fourier sampling:* Quantum algorithms can access the Fourier coefficients of Boolean functions more efficiently than classical algorithms [9, 10]. This underlies exponential speedups for learning specific function classes, such as juntas or parity functions.
- *Quantum kernel estimation:* Classically intractable kernels (e.g. those based on the overlap of quantum states) can be estimated efficiently on a quantum computer, enabling quantum-enhanced classification [11–16].
- *Quantum generative models:* In generative modeling [17], quantum circuits may represent distributions that are difficult to sample from or even describe classically, suggesting a form of expressive advantage [18].

**Key Example:** *Learning DNF formulas.* The class of disjunctive normal form (DNF) formulas is conjectured to be difficult to learn classically in the PAC model, but quantum learners with access to quantum examples can achieve better sample efficiency under certain assumptions [19].

## 1.2 The promise of quantum machine learning

Machine learning has become an indispensable tool across science, technology, and industry, powering advances in areas ranging from medical diagnostics to high energy physics. Despite this success, many learning tasks remain constrained by classical computational resources, in particular when dealing with high-dimensional feature spaces, large-scale generative models, or intractable optimization landscapes. Quantum computing introduces fundamentally different computational principles that offer new possibilities for overcoming these limitations. By harnessing superposition, entanglement, and quantum inference, quantum computers can represent and process information in ways that have no classical analog.

The central promise of QML lies in the hope that quantum algorithms can provide speedups, improved model expressivity, or more efficient data encoding for specific tasks. In some cases, quantum algorithms may reduce the computational complexity of a learning task, enabling solutions to problems that are otherwise infeasible. In others, quantum-enhanced models may explore richer hypothesis spaces or learn data distributions that are difficult to simulate with classical resources. At the interface of these possibilities is a rapidly evolving field that reimagines both the tools and goals of learning theory through the lens of quantum computation.

Importantly, the potential of QML is not limited to achieving runtime improvements. The development of quantum-native models—algorithms that operate directly on quantum data or within quantum systems—suggests new paradigms for information processing. These models are not simply quantum analogs of classical algorithms but often involve novel approaches that exploit the structure of quantum mechanics to redefine what learning can mean. The remainder of this book explores both the theoretical underpinnings and practical implementations of these approaches, with attention to the opportunities and limitations they present for near-term and future quantum hardware.

**When and why quantum models might help**

Quantum models offer potential advantages in machine learning due to their ability to represent, manipulate, and sample from high-dimensional spaces more efficiently than classical models in certain settings. Unlike classical models that operate over real-valued feature vectors, quantum algorithms use complex amplitudes and interference patterns that encode richer geometric structures. QML may help in one or more of the following ways:

- **Computational speedup:** Certain subroutines—such as solving linear systems [20], estimating kernels [11], or sampling from complex distributions—can be executed more efficiently on a quantum computer.
- **Model expressivity:** Quantum circuits can represent probability distributions and function classes that are difficult or impossible for classical models to simulate efficiently [21].
- **Compact data encoding:** Through amplitude [22] or Hamiltonian encoding [15], quantum states can represent exponentially large feature spaces [11, 12], potentially enabling better generalization from fewer physical resources.
- **Quantum advantage on quantum data:** For tasks involving inherently quantum data (e.g. quantum states from sensors or simulators), quantum models offer a natural computational framework with no classical analog [23].

Note that these potential benefits are subject to ongoing research. Not every learning problem benefits from quantum models, and practical performance depends on hardware limitations, the structure of the data, and the design of the algorithm. The goal of this book is to explore where and how such advantages may be realized.

## 1.3 Quantum computing: a refresher

To understand how quantum mechanics can be harnessed for learning tasks, it is helpful to recall the foundational principles of quantum computing. The basic unit of quantum information is the qubit, which, unlike a classical bit, can exist in a superposition of its two basis states. A qubit is described by a complex-valued unit vector in a two-dimensional Hilbert space, and its state evolves through the application of unitary transformations. These unitary operations are implemented using quantum gates, which manipulate individual qubits or pairs of qubits through reversible, deterministic dynamics governed by the Schrödinger equation.

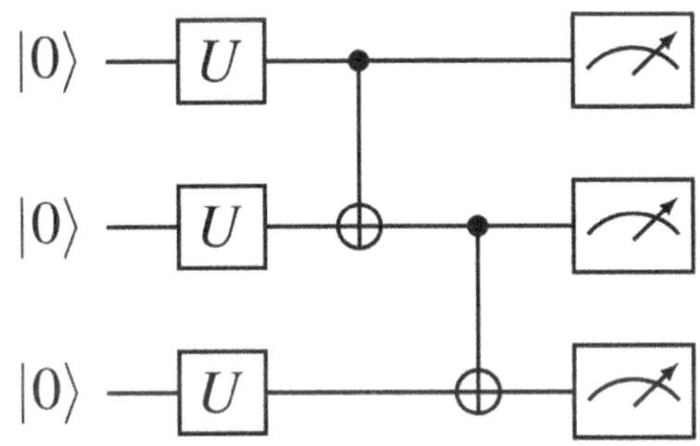


**Figure 1.1.** A simple quantum circuit illustrating the basic structure of quantum computation. The circuit begins with qubit initialization in the $|0\rangle$ state, followed by unitary gate operations (including entangling gates), and ends with measurement in the computational basis. This process forms the core abstraction for quantum algorithms.

An essential feature of quantum systems is entanglement, a form of non-classical correlation that enables the joint state of multiple qubits to encode information that cannot be decomposed into separate components. Entanglement plays a crucial role in many quantum algorithms and is often viewed as a resource that allows for quantum advantage. In practice, quantum computations are carried out by assembling quantum circuits, which consist of a sequence of gate operations acting on an initial quantum state, followed by a final measurement. Measurement projects the quantum state onto a classical outcome according to the Born rule, introducing probabilistic behavior even when the evolution is entirely deterministic (figure 1.1).

Quantum algorithms exploit the unique structure of quantum mechanics to solve problems in ways that are inaccessible to classical computation. The development of such algorithms—such as Grover's search or Shor's factoring—has provided both theoretical insight and practical motivation for building quantum hardware. These foundational algorithms also serve as building blocks in more complex routines, including those used in QML. As we progress through this book, we will revisit these concepts in more detail, particularly in contexts where they directly inform the design of quantum learning models.

## 1.4 Machine learning foundations

Machine learning is a broad and evolving field that aims to develop models capable of recognizing patterns, making predictions, or generating new data by learning from examples. At its core, machine learning involves the approximation of a function or distribution using a dataset sampled from an unknown process. This is typically achieved by selecting a hypothesis class—such as neural networks (NNs)—and optimizing the parameters of a model to minimize some notion of loss or error with respect to training data.

A standard distinction is drawn between supervised and unsupervised learning. In supervised learning, the model is trained on labeled examples, where the goal is to generalize beyond the observed inputs to correctly predict unseen outcomes. In

**Table 1.1.** Correspondence between classical and quantum machine learning terminology. While classical models operate on vectors and real-valued functions, quantum models manipulate quantum states using parameterized circuits and observables.

| Classical machine learning | Quantum machine learning |
|---|---|
| Data | Quantum state |
| Model | Parametric circuit |
| Feature map | Data encoding |
| Loss function | Observable |
| Training | Optimization |

contrast, unsupervised learning seeks to uncover hidden structure in data without explicit labels, often by identifying clusters, reducing dimensionality, or learning generative representations. In both settings, the concepts of generalization, overfitting, and regularization are essential for understanding model performance on unseen data. These challenges arise because the learner must strike a balance between flexibility and inductive bias—choosing a model rich enough to capture relevant patterns but constrained enough to avoid fitting noise.

The training process is typically framed as an optimization problem. A cost function, or loss, quantifies how far the model's predictions are from the desired outcomes, and the model's parameters are adjusted iteratively using optimization algorithms such as gradient descent or its many variants. In modern deep learning, the hypothesis class is often a highly non-linear, layered composition of functions with millions of parameters. See table 1.1 for the correspondence between classical and quantum machine learning terminology. Despite the complexity of such models, empirical evidence shows that they can perform remarkably well when trained on large datasets, provided that the optimization landscape is well-behaved and the model capacity is sufficient.

These foundational ideas will recur throughout the book as we explore quantum models for learning. Many of the concepts—such as model selection, generalization, and optimization—carry over to the quantum domain, but they must be reinterpreted in light of quantum constraints, such as measurement-induced noise, circuit depth limitations, and the probabilistic nature of quantum state preparation and inference. The challenge, and promise, of QML lies in identifying which elements of this classical framework can be enhanced or replaced by quantum mechanisms, and in understanding when such replacements lead to practical or theoretical benefits.

## 1.5 Types of quantum machine learning

QML encompasses a broad range of algorithmic strategies, shaped by the interplay between the data being processed and the computational model used. At its core, QML research asks whether quantum resources can offer an advantage over classical methods in tasks such as classification, regression, sampling, or generative modeling. However, not all QML algorithms engage with quantum mechanics in the same way. The extent to which quantum mechanics is utilized depends not only on whether the algorithm runs on a quantum device, but also on the nature of the input data and how they are represented.

Some QML methods aim to process entirely classical data, using quantum circuits as models with potentially advantageous expressivity or trainability. Others are designed for quantum-native tasks, such as as learning from quantum states produced by physical systems or quantum simulations. Still others fall somewhere in between, using classical data inputs but relying on quantum subroutines—such as state overlap estimation or kernel computation—to accelerate specific components of the learning process. This diversity has led to a need for a coherent framework that distinguishes between the different modes of operation in QML.

To clarify these distinctions, the following taxonomy presents QML approaches in terms of two orthogonal criteria: the type of data being learned from, and the type of model performing the learning. This classification not only provides a conceptual map of the field but also helps highlight the assumptions, limitations, and opportunities inherent in each approach.

**Taxonomy of QML approaches**

QML algorithms can be broadly categorized by two features, the nature of the data and the model/algorithm. This leads to four types of QML workflows:

| Data type | Model type | Example use case |
|---|---|---|
| Classical | Classical | Standard ML: SVMs, CNNs, Gaussian mixture models, gradient boosting |
| Classical | Quantum | Variational quantum classifiers, quantum kernel methods, quantum circuit Born machines trained on classical datasets |
| Quantum | Classical | Shadow tomography, quantum measurement data classification, classical post-processing of quantum sensor outputs |
| Quantum | Quantum | Full quantum generative models, quantum autoencoders, quantum state discrimination, quantum-enhanced metrology |

**Remarks:**

- The *classical data + quantum model* category is most relevant for near-term applications using hybrid quantum–classical algorithms.
- The *quantum data + quantum model* scenario is natural for quantum-native tasks such as quantum error correction, state classification, or entangled sensor networks.
- Some algorithms (e.g. QGANs) can bridge multiple categories depending on how the generator, discriminator, and data encoding are structured.

This chapter established the foundation for exploring QML by outlining the motivations, scope, and structural organization of the book. We introduced the central question driving the field: how quantum computing can enhance or redefine learning tasks in terms of computational efficiency, model expressivity, or data representation. To anchor the reader, we reviewed the key principles of quantum computing and classical machine learning, identifying conceptual parallels and differences that will inform subsequent chapters.

A critical component of this chapter was the classification of QML approaches based on the nature of both the input data and the model, highlighting the spectrum of hybrid and fully quantum strategies that define the field. This taxonomy provides a conceptual map for the reader, helping to locate various algorithms, techniques, and applications within a coherent framework.

By the end of this chapter, the reader should be prepared to navigate the interplay between quantum and classical paradigms, understanding the significance of encoding choices, and appreciate the high-level design space of QML workflows. The remainder of the book builds on this foundation, beginning with a more detailed examination of quantum computational primitives and classical learning theory, and gradually advancing toward the state-of-the-art models, applications, and open challenges in the field.

## References


[1] Biamonte J, Wittek P, Pancotti N, Rebentrost P, Wiebe N and Lloyd S 2017 Quantum machine learning *Nature* **549** 195–202

[2] Montanaro A 2016 Quantum algorithms: an overview *npj Quantum Inf.* **2** 15023

[3] Grover L K 1996 A fast quantum mechanical algorithm for database search *Proc. 28th Annual ACM Symp. on Theory of Computing* pp 212–9

[4] Shor P W 1997 Polynomial-time algorithms for prime factorization and discrete logarithms on a quantum computer *SIAM J. Comput.* **26** 1484–509

[5] Cross A W, Smith G and Smolin J A 2015 Quantum learning robust against noise *Phys. Rev.* A **92** 012327

[6] Ristè D, da Silva M P, Ryan C A, Cross A W, Córcoles A D, Smolin J A, Gambetta J M, Chow J M and Johnson B R 2017 Demonstration of quantum advantage in machine learning *npj Quantum Inf.* **3** 16

[7] Arunachalam S and de Wolf R 2017 Guest column: A survey of quantum learning theory *ACM SIGACT News* **48** 41–67

[8] Servedio R A and Gortler S J 2004 Equivalences and separations between quantum and classical learnability *SIAM J. Comput.* **33** 1067–92

[9] Bernstein E and Vazirani U 1997 Quantum complexity theory *SIAM J. Comput.* **26** 1411–73

[10] Atıcı A and Servedio R A 2007 Quantum algorithms for learning and testing juntas *Quantum Inf. Process.* **6** 323–48

[11] Havlíček V, Córcoles A D, Temme K, Harrow A W, Kandala A, Chow J M and Gambetta J M 2019 Supervised learning with quantum-enhanced feature spaces *Nature* **567** 209–12

[12] Schuld M and Killoran N 2019 Quantum machine learning in feature Hilbert spaces *Phys. Rev. Lett.* **122** 040504

[13] Mitarai K, Negoro M, Kitagawa M and Fujii K 2018 Quantum circuit learning *Phys. Rev.* A **98** 032309
[14] Rebentrost P, Mohseni M and Lloyd S 2014 Quantum support vector machine for big data classification *Phys. Rev. Lett.* **113** 130503
[15] Schuld M, Sweke R and Meyer J J 2021 Effect of data encoding on the expressive power of variational quantum-machine-learning models *Phys. Rev.* A **103** 032430
[16] Kübler J M, Buchholz S and Schölkopf B 2021 The inductive bias of quantum kernels arXiv: 2106.03747
[17] Benedetti M, Garcia-Pintos D, Perdomo O, Leyton-Ortega V, Nam Y and Perdomo-Ortiz A 2019 A generative modeling approach for benchmarking and training shallow quantum circuits *npj Quantum Inf.* **5** 45
[18] Lloyd S and Weedbrook C 2018 Quantum generative adversarial learning *Phys. Rev. Lett.* **121** 040502
[19] Bshouty N H and Jackson J C 1998 Learning DNF over the uniform distribution using a quantum example oracle *SIAM J. Comput.* **28** 1136–53
[20] Harrow A W, Hassidim A and Lloyd S 2009 Quantum algorithm for linear systems of equations *Phys. Rev. Lett.* **103** 150502
[21] Du Y, Hsieh M-H, Liu T and Tao D 2020 Expressive power of parametrized quantum circuits *Phys. Rev. Res.* **2** 033125
[22] Schuld M, Bocharov A, Svore K M and Wiebe N 2020 Circuit-centric quantum classifiers *Phys. Rev.* A **101** 032308
[23] Huang H-Y *et al* 2022 Quantum advantage in learning from experiments *Science* **376** 1182–6

# Chapter 2

# Quantum information processing

Quantum information theory is a branch of physics and computer science that deals with the behavior and manipulation of information at the quantum level. It is based on the principles of quantum mechanics, the fundamental theory of describing the behavior of matter and energy at the atomic and subatomic scales.

In classical information theory, information is represented by bits—fundamental units that can take values of 0 or 1. In contrast, quantum information is represented by **quantum bits**, or **qubits**, which can exist in a superposition of states, enabling more complex and parallel computation.

Quantum information theory has wide applications in quantum computing, quantum communication, quantum cryptography, and quantum metrology. Key concepts include **entanglement, quantum teleportation**, and **quantum error correction**. Entanglement describes correlations between quantum systems that cannot be explained classically. Quantum teleportation allows the transmission of quantum states via classical communication and entanglement. Quantum error correction ensures the robustness of quantum information in the presence of noise.

This chapter introduces core principles of quantum mechanics as they pertain to computation, including qubit states, quantum gates, and circuit-based algorithms. The final section provides an overview of the **postulates of quantum mechanics**, and introduces canonical algorithms such as **Deutsch–Jozsa's** and **Grover's algorithms** to ground future quantum machine learning (QML) discussions.

## 2.1 Quantum computing

Quantum computing uses principles of quantum mechanics—superposition, entanglement, and interference—to perform computations. A **qubit** can be both 0 or a 1 simultaneously due to superposition, and its state is represented in a two-dimensional Hilbert space $\mathscr{H} \cong \mathbb{C}^2$.

doi:10.1088/978-0-7503-4952-9ch2  

A qubit's state $|\psi\rangle = \alpha|0\rangle + \beta|1\rangle$, with $\alpha, \beta \in \mathbb{C}$ and $|\alpha|^2 + |\beta|^2 = 1$, describes a unit vector on the Bloch sphere. Measurement projects the state to $|0\rangle$ or $|1\rangle$ represented by two-dimensional vectors,

$$|0\rangle = \begin{bmatrix} 1 \\ 0 \end{bmatrix}, |1\rangle = \begin{bmatrix} 0 \\ 1 \end{bmatrix}, \tag{2.1}$$

with the probabilities $|\alpha|^2$ and $|\beta|^2$.

Quantum systems also exhibit **entanglement**, a nonlocal correlation that allows multi-qubit states to encode joint properties that classical systems cannot reproduce.

**The quantum bit**

The **quantum bit** or **qubit** is the fundamental unit of quantum information. Unlike a classical bit, which can be found in one of two definite states (0 or 1), a qubit can exist in a **superposition** of both states simultaneously:

$$|\psi\rangle = \alpha|0\rangle + \beta|1\rangle, \quad \text{with} \quad |\alpha|^2 + |\beta|^2 = 1. \tag{2.2}$$

This abstract two-level system can represent a wide range of physical realizations, such as the states $|0\rangle$ and $|1\rangle$ in the computational basis, or $-\frac{1}{2}$ and $\frac{1}{2}$ for a spin$-1/2$ particle.

- The qubit remains in a 'superposition' state until a measurement is performed.
- Measurement collapses the superposition, projecting the qubit into one of the two basis states with probabilities given by $|\alpha|^2$ and $|\beta|^2$.

The behavior of quantum systems is governed by a small set of fundamental rules, known as the **postulates of quantum mechanics** [1, 2]. These postulates define how quantum states are represented, how they evolve in time, how measurement outcomes are determined, and how systems combine.

**Postulates of quantum mechanics**

- **State space:** A quantum system is described by a unit vector $|\psi\rangle$ in a complex Hilbert space.
- **Evolution:** In the absence of measurement, quantum states evolve according to unitary transformations $U$, typically generated by the Schrödinger equation.
- **Measurement:** Observables correspond to Hermitian operators. Upon measurement, the system collapses to an eigenstate of the observable, and the outcome is one of its eigenvalues with probability given by the Born rule.
- **Composite systems:** The state space of a composite system is the tensor product of the component Hilbert spaces. Entanglement arises when the joint state cannot be factored into individual subsystem states.

## 2.2 Dirac notation

Dirac's notation is a formalism used in quantum mechanics to describe quantum states and the time evolution of operators. In Dirac's notation, a vector $\overrightarrow{v}$, is represented as $|v\rangle$. It is based on the use of the *bra–ket* notation, where a **ket** represents a column vector and a **bra** its conjugate transpose. Thus, a qubit that is in the superposition of states 0 and 1, can be represented by the vector or state $|\psi\rangle$. Such state can be represented as the linear combination between $|0\rangle$ and $|1\rangle$ in the following way:

$$|\psi\rangle = \alpha|0\rangle + \beta|1\rangle \tag{2.3}$$

The coefficients $\alpha$ and $\beta$ are complex numbers ($\alpha$, $\beta \in \mathbb{C}$). The *bra* and *ket* represent vectors in a two-dimensional complex Hilbert space where the states $|0\rangle$ and $|1\rangle$ form an orthonormal base. The *bra–ket* notation allows one to represent inner products, outer products, and other operations on these vectors in a concise and intuitive way. For example, in this orthonormal base, we can write $|\psi\rangle$ as

$$|\psi\rangle = \begin{bmatrix} \alpha \\ \beta \end{bmatrix}, \tag{2.4}$$

The physical interpretation of the coefficients $\alpha$ and $\beta$ are the probabilities $|\alpha|^2$ or $|\beta|^2$ of finding our system in state $|0\rangle$ or $|1\rangle$, respectively, after a measurement. Furthermore, these probabilities are normalized to unity. Finally, in Dirac's notation, the bra–ket notation is used to represent *quantum states* and operators. For example, a quantum state $|\psi\rangle$ can be represented as a *ket* and an operator $A$ can be represented as a *bra–ket* pair $\langle\psi|A|\psi\rangle$.

## 2.3 Bloch sphere

The **Bloch sphere** is a graphical representation of the state of a two-level quantum system, such as a qubit. The Bloch sphere is a sphere of unit radius, with the north pole representing the state $|0\rangle$ and the south pole representing the state $|1\rangle$. Any other state of the two-level system can be represented as a point on the surface of the Bloch sphere. For example, the state $|+\rangle$, which is a superposition of $|0\rangle$ and $|1\rangle$, is represented by the point on the equator that is halfway between the north and south poles.

**The Bloch sphere**

Any point in the surface of the Bloch sphere represents a state of the two-level system. The generalization of such a state can be represented through the parameters $\gamma$, $\theta$, $\phi$:

$$|\psi\rangle = e^{i\gamma}\left(\cos\frac{\theta}{2}|0\rangle + \sin\frac{\theta}{2}e^{i\phi}|1\rangle\right). \tag{2.5}$$

The component $e^{i\gamma}$ is a global factor that cannot be detected experimentally and is often omitted, thus reducing the number of parameters needed to describe any point in the Bloch sphere to only two.

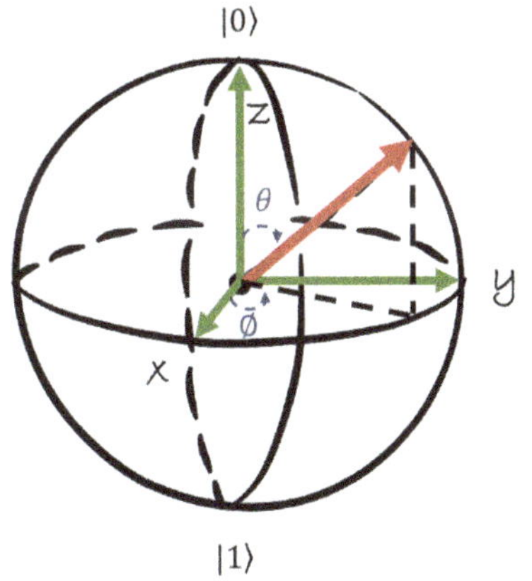


The Bloch sphere is a useful tool for visualizing and understanding the properties of quantum states and the operations that can be performed on them. It is particularly useful for understanding the behavior of quantum gates, which are operations that can be performed on a qubit to change its state. It can also be used to represent the state of a two-level system in the context of quantum error correction, where it can be used to identify and correct errors that occur in the quantum state. In this context, the Bloch sphere is known as the Bloch sphere representation of a quantum error-correcting code.

## 2.4 The Copenhagen interpretation

Although there are infinitely many points on the Bloch sphere, each one corresponding to a possible state for a qubit to occupy, the laws of quantum mechanics dictate that after a measurement, the wave function describing the quantum system collapses to one of the two basis states, $|0\rangle$ or $|1\rangle$. Furthermore, subsequent measurements performed on the system will also yield the same result.

This phenomenon can be understood in the context of the Copenhagen interpretation, developed by Niels Bohr [3] and Werner Heisenberg [4]. This framework for understanding quantum mechanics is based on the idea that the behavior of quantum particles is inherently indeterminate and cannot be predicted with certainty. Instead, the Copenhagen interpretation proposes that the act of measurement or observation causes the quantum state of a system to 'collapse', resulting in a definite outcome.

According to the Copenhagen interpretation [1, 5], quantum systems do not have definite properties until they are measured or observed, and the act of measurement itself determines the outcome. This means that the behavior of quantum particles is fundamentally unpredictable and cannot be described by classical concepts of cause and effect. Instead, it must be described using probability and statistical methods.

Given knowledge of a system's initial state and the sequence of transformations applied to it, the quantum state can be determined precisely at any moment prior to measurement-induced collapse. This principle is the basis for quantum computing. Quantum algorithms can be used to evolve a quantum state from a known initial state to manipulate the probability distribution of the transformed state towards a favorable outcome.

## 2.5 Quantum gates and circuits

In the realm of classical computing, gates are used to manipulate binary data. These gates, such as AND, OR, and NOT, transform input bits into specific output bits based on well-defined logic operations. When transitioning to the quantum realm, we require a new set of tools to perform operations on our quantum data—the qubits. These tools come in the form of quantum gates.

### 2.5.1 Quantum gates

**Quantum gates** are fundamental operations that allow for the manipulation of qubits in a quantum computer. Much like their classical counterparts, these gates dictate the behavior of the system. However, unlike classical gates that deal with definitive 0s and 1s, quantum gates handle qubits, which can be in a superposition of both 0 and 1 states.

The operations of classical gates are deterministic. When you put a certain value, you know exactly what the output will be. Quantum gates, on the other hand, operate in the probabilistic realm of quantum mechanics. Their actions are non-destructive; instead of setting a qubit to a specific state, quantum gates transform qubits in ways that evolve their state in the complex space of possible quantum states. For instance, whereas a classical NOT gate would flip a bit from 0 to 1 or vice versa, its quantum counterpart, the Pauli-$X$ gate, transforms a qubit from the state $|0\rangle$ to $|1\rangle$ and vice versa, but when the qubit is in a superposition state, the outcome becomes less predictable.

Mathematically, quantum gates are represented by unitary matrices. In quantum mechanics, unitary operations are crucial because they preserve the length of vectors in the state space, ensuring that probabilities sum to one. This ensures the conservation of quantum information during computation, aligning with one of the core principles of quantum mechanics. Quantum gates include single-qubit gates ($X$, $Y$, $Z$, $H$, $S$, $T$) and multi-qubit gates (CNOT, Toffoli, SWAP). See table 2.1.

**Universal gate sets**
Any quantum computation can be approximated using the **Clifford + $T$-gate set**. This set combines the Clifford group, generated by the Hadamard (H), Phase (S), and Controlled-Not (CNOT) gates, with the non-Clifford T (or $\pi$ 8) gate. The Clifford gates alone can be efficiently simulated classically, but adding the T gate promotes the set to universality, meaning any unitary operation can be approximated to arbitrary accuracy. The **Solovay–Kitaev theorem** [6] guarantees efficient decomposition into a universal set.

**Table 2.1.** Summary of quantum gates including single, two, and multi-qubit gates.

| Gate name | Matrix representation | Diagram | Description |
|---|---|---|---|
| Pauli-$X$ (NOT) | $\begin{pmatrix} 0 & 1 \\ 1 & 0 \end{pmatrix}$ | $\boxed{X}$ | Bit-flip gate |
| Pauli-$Y$ | $\begin{pmatrix} 0 & -i \\ i & 0 \end{pmatrix}$ | $\boxed{Y}$ | Combines bit and phase flip |
| Pauli-$Z$ | $\begin{pmatrix} 1 & 0 \\ 0 & -1 \end{pmatrix}$ | $\boxed{Z}$ | Phase-flip gate |
| Hadamard ($H$) | $\frac{1}{\sqrt{2}}\begin{pmatrix} 1 & 1 \\ 1 & -1 \end{pmatrix}$ | $\boxed{H}$ | Creates superposition |
| Phase ($S$) | $\begin{pmatrix} 1 & 0 \\ 0 & i \end{pmatrix}$ | $\boxed{S}$ | Introduces phase of $\pi/2$ |
| $T$-gate | $\begin{pmatrix} 1 & 0 \\ 0 & e^{i\pi/4} \end{pmatrix}$ | $\boxed{T}$ | Introduces phase of $\pi/4$ |
| CNOT | $\begin{pmatrix} 1 & 0 & 0 & 0 \\ 0 & 1 & 0 & 0 \\ 0 & 0 & 0 & 1 \\ 0 & 0 & 1 & 0 \end{pmatrix}$ | 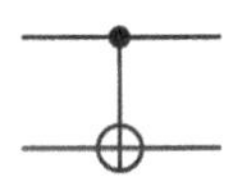 | Controlled-$X$ gate |
| SWAP | $\begin{pmatrix} 1 & 0 & 0 & 0 \\ 0 & 0 & 1 & 0 \\ 0 & 1 & 0 & 0 \\ 0 & 0 & 0 & 1 \end{pmatrix}$ | 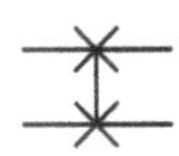 | Swaps two qubits |
| Toffoli (CCX) | $\begin{pmatrix} 1 & 0 & 0 & 0 & 0 & 0 & 0 & 0 \\ 0 & 1 & 0 & 0 & 0 & 0 & 0 & 0 \\ 0 & 0 & 1 & 0 & 0 & 0 & 0 & 0 \\ 0 & 0 & 0 & 1 & 0 & 0 & 0 & 0 \\ 0 & 0 & 0 & 0 & 1 & 0 & 0 & 0 \\ 0 & 0 & 0 & 0 & 0 & 1 & 0 & 0 \\ 0 & 0 & 0 & 0 & 0 & 0 & 1 & 0 \\ 0 & 0 & 0 & 0 & 0 & 0 & 0 & 1 \end{pmatrix}$ | 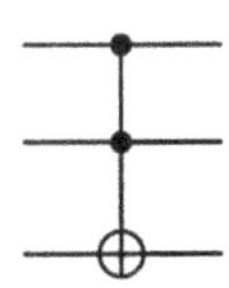 | Controlled-controlled-$X$ gate |

### 2.5.2 The quantum circuit model

A quantum circuit, much like a classical circuit, is a sequence of operations (gates) that performs a specific computation. However, quantum circuits manipulate qubits instead of classical bits. As we dive into the foundational concepts of quantum circuits, it becomes evident that their construction and behavior are inherently distinct and rich compared to classical circuits.

The **quantum circuit model** is the most common framework for designing and analysing quantum algorithms. It provides a visual representation of the quantum operations. In a quantum circuit, qubits are represented as horizontal lines, with time flowing from left to right. Gates acting on the qubits are represented as symbols placed on these lines. When a gate acts on multiple qubits, connecting lines or symbols show the qubits involved. Typically, a quantum circuit starts with all qubits initialized in the $|0\rangle$ state. This standardization simplifies the set-up, but it is essential to understand that quantum algorithms can use and produce any superposition of states. Once the initial state is set, the computation begins by applying a series of quantum gates. These gates, which can act on one, two, or more qubits, transform the qubits' states. Notably, while a single-qubit gate only affects one qubit, a two-qubit gate can entangle the qubits, creating a link between them that is fundamental to many quantum algorithms.

After applying a series of quantum gates, a measurement is often performed on one or more qubits. In classical computing, reading the state of a bit is straightforward. In contrast, measuring a qubit is more complex due to the principles of quantum mechanics. A measurement collapses the qubit's state to either $|0\rangle$ or $|1\rangle$, with probabilities determined by the state's amplitude squared. The outcome is inherently probabilistic, and if the same quantum circuit is run multiple times, the measurement might yield different results each time.

An essential feature of quantum circuits is the ability to perform computations in parallel thanks to superposition. By placing a qubit in a superposition of both $|0\rangle$ and $|1\rangle$ states, the circuit can compute both states simultaneously. This trait does not mean that quantum computers can solve all problems faster than classical ones, but it does open the door for specific algorithms where this parallelism can be exploited for speedups. Finally, when qubits are entangled, the state of one qubit is dependent on the state of another. This powerful feature allows quantum circuits to represent and manipulate a vast amount of information in ways that classical circuits cannot.

Grasping the basics of quantum circuits is pivotal for understanding quantum computing as a whole. These circuits are the playground where quantum principles come alive, showcasing the peculiarities and immense potential of quantum mechanics in computation. As one grows familiar with these basics, they serve as a stepping stone to more advanced quantum algorithms and applications.

## 2.6 The Deutsch–Jozsa algorithm

The Deutsch–Jozsa algorithm [7] was the first quantum algorithm to demonstrate a clear advantage over classical computation. It solves a problem that is exponential in the worst-case classical setting using only a single evaluation on a quantum computer.

### 2.6.1 Problem statement

Let $f: \{0, 1\}^n \rightarrow \{0, 1\}$ be a Boolean function that is promised to be either constant (the same output for all inputs) or balanced (outputs 0 for exactly half the inputs and 1 for the other half). The task is to determine which case applies.

Classically, in the worst case, one must evaluate $f$ on $2^{n-1} + 1$ inputs. The Deutsch–Jozsa algorithm solves this with a single invocation of a quantum oracle (see figure 2.1).

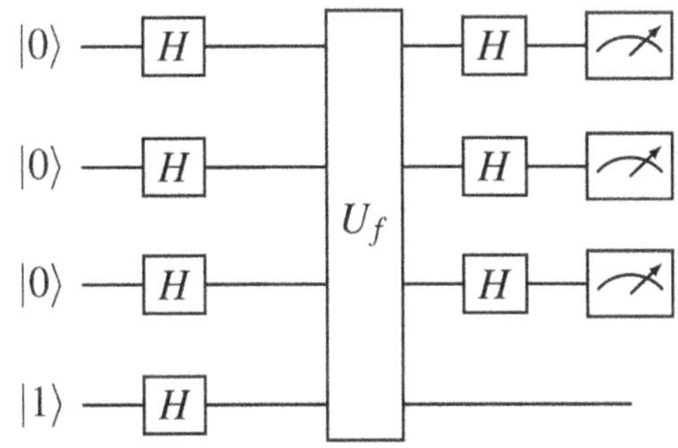


**Figure 2.1.** Quantum circuit for the Deutsch–Jozsa algorithm with three input qubits and one ancilla qubit. The input register is initialized in the $|0\rangle^{\otimes 3}$ state and the ancilla in $|1\rangle$. After applying Hadamard gates to all qubits, the oracle $U_f$ encodes the function $f(x)$ into the phase of the quantum state. A second round of Hadamard gates is applied to the input register, followed by measurement. The outcome distinguishes between constant and balanced functions in a single query.

### 2.6.2 Algorithm overview

1. **Initialize** an $(n + 1)$-qubit system in the state $|0\rangle^{\otimes n} \otimes |1\rangle$.
2. **Apply Hadamard gates** to all $n + 1$ qubits to create a superposition.
3. **Invoke the oracle** $U_f$, which maps $|x\rangle|y\rangle \mapsto |x\rangle|y \oplus f(x)\rangle$.
4. **Apply Hadamard gates** to the first $n$ qubits.
5. **Measure** the first $n$ qubits.

### 2.6.3 Interpretation of results

- If the output is $|0\rangle^{\otimes n}$, then $f$ is constant.
- Otherwise, $f$ is balanced.

### 2.6.4 Significance

The algorithm showcases the power of **quantum parallelism** and **interference**. By evaluating all inputs in superposition and interfering the amplitudes constructively or destructively, the Deutsch–Jozsa algorithm achieves exponential speedup in query complexity.

## 2.7 Grover's algorithm

Grover's algorithm [8] provides a quantum solution to the unstructured search problem, offering a **quadratic speedup** over classical brute-force methods.

### 2.7.1 Problem statement

Given an unstructured database of $N = 2^n$ items and an oracle function $f$: $\{0, 1\}^n \to \{0, 1\}$ that returns 1 only for the target item and 0 otherwise, find an input $x_0$ such that $f(x_0) = 1$. Classically this requires $O(N)$ evaluations. Grover's algorithm achieves this in approximately $O(\sqrt{N})$ steps (see figure 2.2).

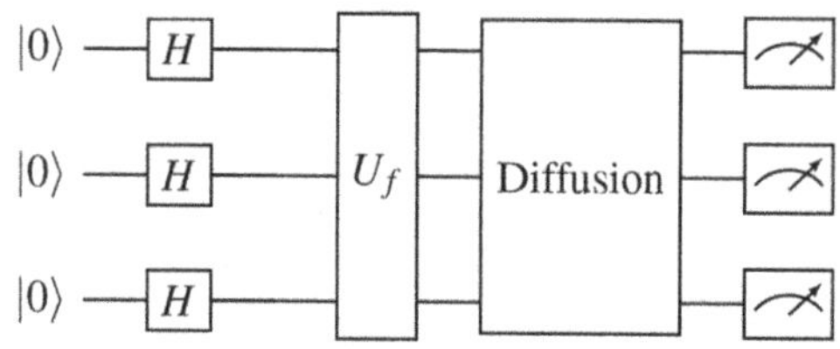


**Figure 2.2.** Quantum circuit for Grover's algorithm with three qubits. The initial Hadamard gates prepare a uniform superposition over all possible inputs. The oracle $U_f$ flips the phase of the marked state. The diffusion operation (also known as the inversion-about-the-mean) amplifies the probability amplitude of the marked state. After one iteration of oracle and diffusion, measurement reveals the target state with high probability.

In the Deutsch–Jozsa algorithm, the oracle $U_f$ encodes a *black-box function* $f$: $\{0, 1\}^n \to \{0, 1\}$. Physically, it is a unitary operation that queries this function coherently, meaning it evaluates $f(x)$ on all possible inputs $x$ simultaneously due to quantum superposition. The oracle acts on two quantum registers according to

$$|x\rangle|y\rangle \mapsto |x\rangle|y \oplus f(x)\rangle, \tag{2.6}$$

where $x$ is the $n-$qubit input register that holds the argument of the function, and $y$ is a one-qubit target register that stores the result. The symbol $\oplus$ denotes addition modulo 2.

In this mapping, the first register $|x\rangle$ remains unchanged, while the second register is flipped if and only is $f(x) = 1$. This controlled flip can be interpreted as a reversible implementation of a classical logic gate. For example, when $f(x)$ corresponds to a Boolean expression, $U_f$ acts like a quantum version of a classical circuit that computes $f$ without destroying superpositions.

This means that $U_f$ coherently 'writes' the function value $f(x)$ into the phase or amplitude of the quantum state, allowing interference patterns to later reveal global properties of $f$.

### 2.7.2 Algorithm overview

1. **Initialize** an $n$-qubit system in the state $|0\rangle^{\otimes n}$ and apply Hadamard gates to create an equal superposition.
2. **Repeat** the Grover iteration $O(\sqrt{N})$ times:
   - **Oracle step:** Apply $U_f$, which flips the phase of the marked item.
   - **Diffusion operator:** Reflect the state about the average amplitude.
3. **Measure** the final state to obtain $x_0$ with high probability.

### 2.7.3 Mathematical insight

Let $|s\rangle = \frac{1}{\sqrt{N}}\sum_{x=0}^{N-1}|x\rangle$ be the initial superposition. Grover's iteration amplifies the amplitude of the marked state by rotating the quantum state toward $|x_0\rangle$ in a two-dimensional subspace spanned by $|x_0\rangle$ and $|s\rangle$.

### 2.7.4 Significance

Grover's algorithm is optimal in the quantum setting for unstructured search, demonstrating a quadratic speedup. It is widely applicable to optimization, satisfiability, and machine learning tasks where solution checking is easy but finding the solution is hard.

## 2.8 Summary

This chapter introduced the foundational elements of quantum information processing, essential for understanding quantum machine learning. We began by contrasting classical and quantum representations of information, emphasizing the role of qubits and their unique properties—superposition, entanglement, and measurement.

The formal structure of quantum theory was introduced through Dirac notation, the Bloch sphere, and the postulates of quantum mechanics. These postulates govern the behavior of quantum systems, including how states evolve and how measurement outcomes are obtained.

Building on these concepts, we explored the quantum circuit model and key quantum gates, establishing a language for designing quantum algorithms. The chapter culminated with detailed presentations of two canonical quantum algorithms: the Deutsch–Jozsa algorithm, which offers an exponential speedup for promise problems, and Grover's algorithm, which provides a quadratic speedup for unstructured searches.

Together, these topics provide the essential computational primitives and conceptual foundations for the quantum algorithms and machine learning models discussed in the remainder of the book.

## References

[1] Sakurai J J and Napolitano J 2010 *Modern Quantum Mechanic* 2nd edn (San Francisco, CA: Addison-Wesley)

[2] Ballentine L E 1998 *Quantum Mechanics: A Modern Development.* (Singapore: World Scientific)

[3] Bohr N 1987 *The Philosophical Writings of Niels Bohr, Vol. I–III* **vols 1–3** (Woodbridge, CT: Ox Bow Press)

[4] Heisenberg W 1958 *Physics and Philosophy: The Revolution in Modern Science.* (New York: Harper and Brothers)

[5] Jammer M 1974 *The Philosophy of Quantum Mechanics* (New York: Wiley)

[6] Kitaev A Y 1997 Quantum computations: algorithms and error correction *Russ. Math. Surv.* **52** 1191–249

[7] Deutsch D and Jozsa R 1992 Rapid solution of problems by quantum computation *Proc. R. Soc. Lond.* A **439** 553–8

[8] Grover L K 1996 A fast quantum mechanical algorithm for database search *Proc. 28th Annual ACM Symp. on Theory of Computing* pp 212–9

# Chapter 3

## Information encoding

Encoding classical data into quantum states is a foundational task in quantum computing and, by extension, quantum machine learning (QML). The choice of encoding not only determines how classical information is mapped to the Hilbert space but also directly influences the expressiveness and trainability of quantum circuits, which in turn influences the efficiency of quantum algorithms. The study of quantum data encoding can be traced back to early quantum algorithms such as the quantum Fourier transform (QFT) [1] and Grover's algorithm, where classical inputs were embedded using basis encoding or amplitude encoding techniques.

The theoretical implications of data encoding gained renewed attention as QML matured into a distinct field. Schuld and Petruccione (2018) [2], and later Schuld, Bocharov, Wiebe, and Svore (2020) [3], emphasized that the data-loading step can be computationally costly and may negate quantum speedups if not addressed carefully. Various encoding strategies have since been proposed, including basis, amplitude, angle, and Hamiltonian based encodings, each with trade-offs in terms of circuit depth, qubit count, and noise resilience.

Recent literature has expanded to consider the inductive bias introduced by different encodings [4], the complexity of state preparation [5], and the limitations posed by barren plateaus during training [6]. A recurring theme is that while quantum models offer theoretical advantages, their practical utility hinges on encoding schemes that are both expressive and resource-efficient.

This chapter provides a comprehensive treatment of information encoding strategies for QML. We begin by reviewing classical approaches to data representation and proceed to formalize the quantum counterparts using rigorous mathematical formulations. We then identify major challenges associated with encoding, including state preparation complexity, hardware limitations, and optimization issues. Finally, we highlight emerging opportunities in the design of encoding strategies for noisy intermediate-scale quantum (NISQ)-era algorithms and beyond.

doi:10.1088/978-0-7503-4952-9ch3  

## 3.1 Bridging two computational realms

In the realm where classical data interfaces with quantum systems, the linchpin that facilitates the flow of information is classical-to-quantum encoding. This essential conduit allows quantum processors, with their inherently quantum nature, to comprehend and manipulate data that originate in the classical domain. In this section, we will delve deep into the intricacies of this conversion process, exploring various methods and considering their implications.

In classical machine learning, data representation plays a central role in determining the performance of learning algorithms. Consider a dataset $\mathscr{D} = \{(x_i, y_i)\}_{i=1}^{N}$, where $x_i \in \mathbb{R}^d$ are input features and $y_i \in \mathbb{R}$ the corresponding labels. A common approach in QML is to transform the input space using a **quantum feature map** [7]$\phi$: $\mathbb{R}^d \to \mathscr{F}$, where $\mathscr{F}$ is a high-dimensional feature space, often a Hilbert space. This transformation enables the application of linear methods in $\mathscr{F}$ to capture nonlinear relationships in the original input space.

A classical analog to quantum encoding is found in kernel methods, where data are mapped implicitly through a kernel function $k(x, x') = \langle\phi(x), \phi(x')\rangle$. Popular feature maps include polynomial and radial basis functions (RBFs), which effectively encode prior assumptions about the structure of the data and the task.

**Feature maps**

Consider a Hilbert space denoted as $\mathscr{H}$, referred to as the feature space. Let $\chi$ represent an input set, and $x$ represent an individual sample from this input set. In this context, a **feature map** is a function, denoted as $\phi$, that operates on inputs from the set $\chi$ and maps them to vectors within the Hilbert space $\mathscr{H}$,

$$\phi: \chi \to \mathscr{H},$$

such that

$$\phi(x) = U(x)|0\rangle^{\otimes n}.$$

Then a quantum model constructs a function

$$f(x) = \langle 0 |^{\otimes n} U^{\dagger}(x) M U(x)|0\rangle^{\otimes n}, \tag{3.1}$$

where $M$ is a Hermitian observable. This representation allows for kernel-based approaches via the quantum kernel

$$k(x, x') = ||\langle\phi(x)|\phi(x')\rangle||^2, \tag{3.2}$$

which has been studied in the context of quantum support vector machines (QSVMs) [8] and other learning models.

By employing a series of quantum gates to induce superpositions and entanglements, quantum feature maps generate complex quantum states that effectively represent the underlying data. These states find applications in quantum kernel methods and various other QML algorithms.

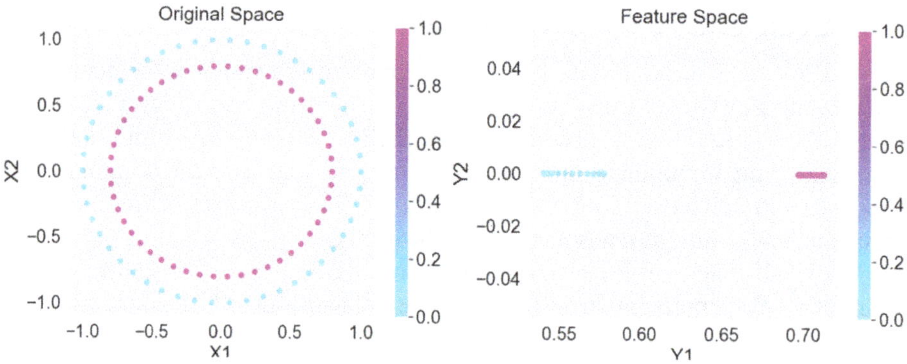


**Figure 3.1.** An example of non-separable data. We want to translate our original space to Hilbert space. The feature map is designed to make non-separable points in the original space separable in the feature space. This idea is very close to the kernel-trick in a classical SVM.

As an example, in figure 3.1, we showcase the effect of applying a second-order quantum feature map to a non-separable dataset consisting of two concentric circles (figure 3.1 (left)). The quantum feature map $U_{\phi(\mathbf{x})}$ contains three operators $U1_{\phi(x_1)}$, $U1_{\phi(x_2)}$, and $U1_{\phi(x_1,\, x_2)}$ mixed with Hadamard gates and CNOT gates. Here $U1$ is a phase shifting gate,

$$U1_{\phi} = \begin{bmatrix} 1 & 0 \\ 0 & e^{i\phi} \end{bmatrix}, \tag{3.3}$$

and the function $\phi$ is defined as

$$\phi = \begin{cases} \phi(x_{1,2}) = x_{1,2}, \\ \phi(x_1,\, x_2) = \pi x_1 x_2 \end{cases}. \tag{3.4}$$

We then measure the value of $\sigma^z\sigma^z$ operator, where $\sigma^z$ is the Pauli $Z$ operator.

$$\sigma^z\sigma^z = \mathbf{ZZ} = \begin{bmatrix} 1 & 0 \\ 0 & -1 \end{bmatrix} \otimes \begin{bmatrix} 1 & 0 \\ 0 & -1 \end{bmatrix} = \begin{bmatrix} 1 & 0 & 0 & 0 \\ 0 & -1 & 0 & 0 \\ 0 & 0 & -1 & 0 \\ 0 & 0 & 0 & 1 \end{bmatrix}. \tag{3.5}$$

After applying the feature map, we can see that datapoints are elevated to a feature space where they are separable (figure 3.1, right). However, it is important to note that designing efficient and useful feature maps can be a formidable task. Overly complex maps may also pose challenges during the subsequent measurement stage when extracting valuable information.

Another intriguing facet is the concept of **quantum embeddings**, which aim to provide a deep representation of classical data. Similar to deep learning embeddings that capture intricate data patterns, quantum embeddings seek to represent classical data using quantum states. This approach preserves the inherent relationships and structures within the data. Typically, data are first transformed using classical

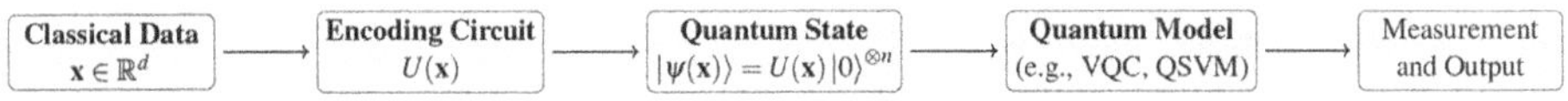


**Figure 3.2.** End-to-end quantum machine learning workflow involving classical data encoding. Classical data vectors are mapped to quantum states via parameterized encoding circuits. The resulting states are processed by a quantum model, and the outcome is obtained through measurement.

techniques and then inputted into quantum circuits, which generate specific quantum states. This process often involves methods such as amplitude and phase encoding. However, it is worth noting that, akin to deep learning embeddings, quantum embeddings may at times obscure interpretability, making it difficult to comprehend why certain predictions or classifications are made.

The choice of encoding method largely depends on the nature of the data and the specific quantum machine learning task at hand. Each approach brings its own set of advantages and challenges, and understanding their nuances is crucial for effectively harnessing the power of quantum systems in classical data processing and analysis. In figure 3.2, a general workflow for QML involving classical data is displayed.

## 3.2 Encoding schemes for quantum machine learning

When we dive into the vast ocean of QML, one of the essential aspects we encounter is the various encoding schemes tailored for QML. These schemes play a pivotal role in determining how effectively and efficiently classical data can be processed using quantum algorithms.

In the quantum setting, a classical vector $x \in \mathbb{R}^d$ is embedded into a quantum state $|\psi(x)\rangle \in \mathscr{H}^{\otimes n}$, where $\mathscr{H} \cong \mathbb{C}^2$ denotes the Hilbert space of a single qubit, and $n$ is the number of qubits. This embedding is typically implemented via a parameterized unitary operator $U(x)$, acting on the all-zero initial state: $|\psi(x)\rangle = U(x)|0\rangle^{\otimes n}$. We outline several common encoding strategies below.

### 3.2.1 Amplitude encoding

At its core, amplitude encoding revolves around using the amplitude of quantum states to represent data. Imagine a grayscale image where pixel intensities vary between black and white. In amplitude encoding, each intensity level can be mapped to a specific amplitude of a quantum state. This allows for a highly compact representation, in particular for data with a lot of variables, as the values of these variables can be represented as normalized amplitudes in a quantum state.

Amplitude encoding uses the components of a normalized real-valued vector $x \in \mathbb{R}^d$, with $\|x\| = 1$, to specify the amplitudes of a quantum state:

$$|x\rangle = \sum\nolimits_{i=1}^{d} x_i |i\rangle . \tag{3.6}$$

This encoding requires only $n = \log_2 d$ qubits, but preparing such a state generally involves a circuit of depth $\mathcal{O}(d)$, potentially requiring ancilla qubits and nontrivial gate synthesis.

### 3.2.2 Angle/parametric encoding

This method takes a different approach, using the relative phase between quantum states to hold information. In essence, data values are encoded in the angle of the quantum state's phase, allowing quantum algorithms to then manipulate these angles. It is akin to encoding information in the direction the quantum state points in a complex plane, providing another avenue for representing continuous data values.

Angle encoding maps each component of $x$ to the rotation angle of a single-qubit gate:

$$x_i \mapsto R_\alpha(x_i) = \exp\left(-i\frac{x_i}{2}\sigma^\alpha\right), \quad \alpha \in \{x, y, z\}, \tag{3.7}$$

where $\sigma^\alpha$ denotes the corresponding Pauli matrix. Angle encoding is hardware-friendly and particularly suited for variational quantum algorithms implemented on noisy intermediate-scale quantum (NISQ) devices.

### 3.2.3 Basis encoding

In basis encoding, each classical data point corresponds to a particular quantum state in the computational basis. For instance, classical binary data such as 01 can be mapped directly to the quantum state $|01\rangle$:

$$x \in \{0, 1\}^n \mapsto |x\rangle = |x_1\rangle \otimes \cdots \otimes |x_n\rangle . \tag{3.8}$$

This method provides a straightforward way to store classical information, in particular for data that already exist in binary form. This approach is efficient in qubit usage for binary strings but lacks expressivity for general learning tasks.

### 3.2.4 Hamiltonian encoding

Hamiltonian encoding refers to a class of quantum data embedding strategies where classical or structured input data ($x$) is used to define a Hamiltonian ($H(x)$), and the quantum state is prepared through its unitary evolution. Formally, given a parameter-dependent Hamiltonian ($H(x)$), the encoding is implemented via the unitary operator $U(x) = e^{-iH(x)t}$ applied to some fixed initial state ($|0\rangle^{\otimes n}$) or another reference state. The result is a quantum state ($|\psi(x)\rangle = U(x)|0\rangle^{\otimes n}$) that encodes the data implicitly in its time-evolved dynamics. This approach is sometimes referred to as *Hamiltonian simulation-based encoding*.

Hamiltonian encoding is particularly well-suited for structured data with a natural physical interpretation, such as graphs, spin systems, or molecular structures, where one can construct a meaningful Hamiltonian representation of the input. It is also used in quantum kernel methods, where the feature map corresponds to the evolution under data-dependent generators.

A common choice in practice is to decompose the Hamiltonian into a sum of Pauli terms, $H(x) = \sum_j x_j P_j$, where ($x_j$) are real-valued coefficients derived from the classical input and ($P_j$) are tensor products of Pauli operators (i.e., elements of the Pauli group). This form allows for efficient implementation of ($U(x)$) using

Trotter-Suzuki decompositions or product formulas, especially when the terms in the Hamiltonian commute or can be grouped into commuting subsets.

**Example: graph encoding.**

Consider a graph $(G = (V, E))$ with adjacency matrix (A). One can define a Hamiltonian $H_G = \sum_{(i,j)\in E} Z_i Z_j$, where $(Z_i)$ is the Pauli-(Z) operator acting on qubit (i). Then the encoding becomes $U_G = e^{-i\theta H_G} = \exp\left(-i\theta \sum_{(i,j)\in E} Z_i Z_j\right)$, and the graph structure is embedded into the entanglement pattern of the evolved quantum state. Such encodings are useful in graph-based learning tasks or in modeling energy-based systems.

**Properties.**

- **Data type:** Structured data (e.g., graphs, molecules), or real-valued vectors mapped to Pauli coefficients.
- **Qubit requirement:** Depends on the system size (n); typically moderate to high.
- **Circuit depth:** Variable; depends on the Hamiltonian and simulation method (e.g., Trotterization, hardware-native gates).
- **Noise resilience:** Moderate. While expressivity is high, long simulation times or deep circuits can introduce decoherence.
- **Expressivity:** High. Hamiltonian encodings can represent complex correlations and entanglement patterns.

**Applications.**

Hamiltonian encoding has found applications in quantum kernel methods (e.g., data re-uploading models), quantum generative learning, and phase classification tasks. It also naturally arises in quantum simulation workflows, where learning is applied to time-evolved or ground-state data generated by physical models. Table 3.1 compares the major encoding schemes.

**Table 3.1.** Comparison of common quantum data encoding strategies. Each strategy balances different trade-offs in terms of qubit requirements, circuit complexity, noise resilience, and applicability to data types.

| Encoding strategy | Data type | Qubit requirement | Circuit depth | Noise resilience | Example gates |
|---|---|---|---|---|---|
| Basis encoding | Binary strings | $n$ | Low | High | $X, I$ |
| Angle encoding | Real-valued features | $n$ | Low–medium | Moderate | $R_y(x_i)$, $R_z(x_i)$ |
| Amplitude encoding | Normalized real vector $x \in \mathbb{R}^d$ | $\log_2 d$ | High | Low | State preparation circuit |
| Hamiltonian-based encoding | Structured inputs (e.g. graphs, molecules) | Problem-dependent | Variable | Moderate | $e^{-iH(x)t}$ |

## 3.3 Quantum autoencoders

Autoencoders [9, 10] are a class of machine learning models designed to learn efficient, lower-dimensional representations of data. In the classical setting, an autoencoder consists of an encoder $E\colon \mathbb{R}^d \to \mathbb{R}^{d'}$ and a decoder $D\colon \mathbb{R}^{d'} \to \mathbb{R}^d$, trained jointly to minimize reconstruction loss:

$$\mathscr{L}(x) = \|D(E(x)) - x\|^2. \tag{3.9}$$

The learned latent representation $z = E(x)$ ideally captures the most salient features of the input data in a compressed form.

Quantum autoencoders (QAEs) [11] extend this concept to quantum data (see table 3.2 for a comparison of classical and quantum autoencoders), aiming to compress quantum states into a smaller Hilbert space while preserving the essential information. Let $|\psi\rangle \in \mathscr{H}_{\text{in}} = \mathscr{H}_A \otimes \mathscr{H}_B$, where $\dim(\mathscr{H}_A) = 2^n$ and $\dim(\mathscr{H}_B) = 2^m$. A QAE seeks a unitary $U$ such that

$$U|\psi\rangle \approx |\phi\rangle_A \otimes |0\rangle_B, \tag{3.10}$$

where $|\phi\rangle_A \in \mathscr{H}_A$ encodes the compressed information and $|0\rangle_B$ is a fixed reference state of the discarded subsystem.

The loss function used in variational quantum autoencoders (VQAEs) [12] is typically of the form

$$\mathscr{L}_{\text{QAE}} = 1 - \text{Tr}[\rho_B|0\rangle\langle 0|], \tag{3.11}$$

where $\rho_B = \text{Tr}_A[U|\psi\rangle\langle\psi|U^\dagger]$ is the reduced density matrix of the discarded subsystem. This loss is minimized when the discarded subsystem is disentangled and in the reference state, allowing for successful compression.

### 3.3.1 Circuit construction

A VQAE is implemented via a parameterized quantum circuit $U(\theta)$ trained over a set of input states $\{|\psi_i\rangle\}$. After optimization, $U(\theta^*)$ encodes the essential degrees of

**Table 3.2.** Comparison of classical and quantum autoencoders.

| Aspect | Classical autoencoder | Quantum autoencoder |
|---|---|---|
| Input | Real vector $x \in \mathbb{R}^d$ | Quantum state $|\psi\rangle \in \mathscr{H}_{\text{in}}$ |
| Latent space | $\mathbb{R}^{d'}$ with $d' < d$ | Subsystem $\mathscr{H}_A \subset \mathscr{H}_{\text{in}}$ |
| Encoder | Neural network $E(x)$ | Parameterized unitary $U(\theta)$ |
| Decoder | Neural network $D(z)$ | $U^\dagger(\theta)$ |
| Loss function | Reconstruction error $\|D(E(x)) - x\|^2$ | Projection fidelity $1 - \text{Tr}[\rho_B|0\rangle\langle 0|]$ |
| Applications | Dimensionality reduction, denoising | State compression, error mitigation, quantum channel capacity |

freedom into a subset of qubits while mapping the remaining qubits to $|0\rangle$. The decoder is given by $U^{\dagger}(\theta^{*})$, which recovers an approximation of the original state:

$$|\psi_i\rangle \approx U^{\dagger}(\theta^{*})(|\phi_i\rangle_A \otimes |0\rangle_B). \tag{3.12}$$

### 3.3.2 Applications and relevance

QAEs are a powerful and flexible component within the broader framework of QML. At their core, QAEs are designed to compress quantum states into a smaller subspace while preserving the essential information needed for downstream tasks. This ability to reduce the dimensionality of quantum states has far-reaching implications for both practical applications and theoretical investigations. In contexts where quantum resources such as memory, coherence time, or qubit availability are limited, QAEs offer a route toward more efficient use of these scarce resources. They enable the representation of complex quantum data using fewer qubits, which is especially valuable for near-term quantum devices operating under noise and hardware constraints.

From a physical sciences perspective, QAEs can be applied to compress quantum states arising from quantum chemistry, condensed matter, or field-theoretic simulations. By identifying a reduced subspace sufficient to represent a family of related quantum states, QAEs can accelerate simulation pipelines and offer insight into the underlying structure of the data manifold [11]. In the context of quantum communication, QAEs can serve as tools to approximate the minimal dimension required to faithfully transmit or store quantum information, thereby providing estimates of quantum channel capacity [13].

In QML, QAEs serve as the natural analog of classical autoencoders, offering pathways for unsupervised representation learning and quantum generative modeling. The latent representations learned by QAEs can support classification [14] and anomaly detection [15] tasks. Moreover, QAEs have been proposed as mechanisms for quantum error mitigation, particularly by identifying and projecting onto decoherence-free subspaces or noiseless subsystems [16, 17]. This function aligns them closely with broader goals in fault-tolerant computation. As quantum hardware continues to evolve, QAEs are expected to play an increasingly important role as modular components in hybrid quantum–classical workflows, integrated with variational circuits, quantum neural networks, and generative models.

## 3.4 Quantum data

As the world of quantum computing expands its horizons, there is a growing interest in not just processing classical data with quantum systems, but in understanding and processing inherently quantum data. This fusion of quantum mechanics and machine learning creates an avenue where data are not just a binary bit or a continuous value, but something intrinsically quantum, often represented by complex quantum states.

The notion of *quantum data* refers to information that is inherently quantum in nature, represented by quantum states rather than classical bit strings. Unlike classical data which are composed of deterministic or probabilistic assignments over discrete variables, quantum data are encoded in the state of a quantum system, described mathematically by a vector in a complex Hilbert space (for pure states) or density matrix (for mixed states). Quantum data arise naturally in a variety of scientific domains, including quantum physics, chemistry, metrology, and quantum information processing itself.

### 3.4.1 Definition and representations

Formally, quantum data correspond to a state $\rho \in \mathscr{D}(\mathscr{H})$, where $\mathscr{D}(\mathscr{H})$ denotes the set of density operators on a Hilbert space $\mathscr{H}$. Pure quantum data are described by $\rho = |\psi\rangle\langle\psi|$, with $|\psi\rangle \in \mathscr{H}$, while mixed quantum data may represent classical uncertainty over quantum states, or entanglement with an inaccessible environment. These states may be prepared experimentally or arise as outputs of quantum circuits or physical simulations.

From an information-theoretic perspective, quantum data differ fundamentally from classical data due to key properties such as:

- **Superposition:** A quantum state can encode a continuum of amplitudes, leading to richer representational capacity.
- **Entanglement:** Quantum data may exhibit non-classical correlations across subsystems, which can encode structured global information.
- **No-cloning:** Quantum data cannot, in general, be copied without disturbance, limiting access to training examples or labeled data in a supervised learning context.
- **Measurement-induced collapse:** Extracting classical information from quantum data is fundamentally probabilistic and destructive.

### 3.4.2 Sources of quantum data

Quantum data may be generated from a number of sources, depending on the application domain (figure 3.3):

- **Simulated quantum systems:** Quantum states from condensed matter, lattice gauge theory, or molecular systems simulated via classical or quantum algorithms.
- **Experimental systems:** Data collected from quantum optics, superconducting circuits, trapped ions, or other quantum hardware.
- **Intermediate quantum circuits:** Quantum states produced at intermediate stages of quantum algorithms, such as ansatzes in variational quantum algorithms or latent states in generative models.
- **Sensor networks and metrology:** Quantum states encoding physical parameters of interest, such as fields, forces, or time intervals, used in precision measurement tasks.

In the context of quantum machine learning, it is crucial to distinguish between algorithms that process *classical data using quantum models* and those that process *quantum data natively*. The latter setting is more general, but also presents greater challenges in terms of state preparation, access to labels, and validation.

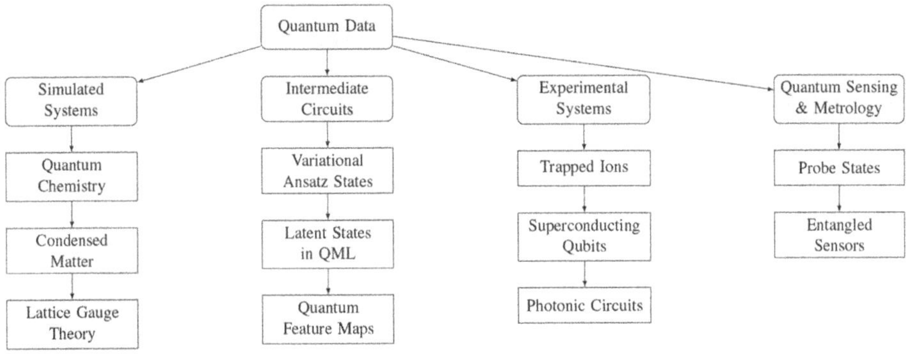


**Figure 3.3.** Sources and types of quantum data relevant to quantum machine learning. Quantum data may originate from simulated physical systems, intermediate circuit states, experimental hardware, or quantum-enhanced sensing devices.

### 3.4.3 Quantum learning from quantum data

Learning from quantum data involves estimating functions, decision boundaries, or generative models from a training set $\{\rho_i, y_i\}$, where $\rho_i$ are quantum states and $y_i$ are classical or quantum labels. Several settings have been explored:

- **Quantum classification:** Determine a label $y$ from an unknown quantum state $\rho$ via a quantum measurement strategy or variational circuit [18–20].
- **Quantum regression:** Learn a mapping from quantum states to real-valued observables or expectation values [21].
- **Quantum generative modeling:** Learn to generate a distribution over quantum states or to reproduce the statistics of quantum data [22].

One key challenge in this regime is access: quantum data cannot be freely copied and are often available only through quantum oracles or experimental runs. Techniques such as quantum state tomography are limited in scalability [23, 24], prompting the need for *informationally efficient learning methods* that require few copies or indirect observables. Recent work has also explored learning quantum measurements, i.e. optimizing positive operator-valued measures (POVMs) [25]—a measure whose values are positive semi-definite operators in a Hilbert space.

### 3.4.4 Implications for encoding

Encoding quantum data is conceptually different from encoding classical data. In the classical case, data must first be embedded into a quantum state via unitary transformations or state preparation routines. In contrast, for quantum data, the challenge lies in designing quantum machine learning models that preserve and exploit the structure of the input quantum state, often without destructive measurement.

Furthermore, standard assumptions in classical learning theory—such as i.i.d. sampling or unrestricted access to data—do not carry over straightforwardly to quantum settings. New theoretical frameworks, such as PAC learning with quantum examples, quantum statistical learning, and quantum generalization bounds, are actively being developed to bridge this gap.

## 3.5 Practical considerations for quantum encoding

In the landscape of quantum computing, theoretical elegance must be carefully balanced with the constraints of practical implementation. Quantum encoding, the intricate process of translating classical data into the quantum domain, is a particularly sensitive point of intersection between these two domains. While various encoding strategies offer distinct advantages in terms of expressivity and performance, their effectiveness ultimately depends on how well they align with the limitations of current quantum hardware and the nature of the input data. One of the most immediate constraints stems from **hardware limitations**. Contemporary quantum processors, often referred to as NISQ devices, provide only a limited number of qubits and suffer from relatively high error rates. An encoding scheme that is theoretically optimal may nonetheless be infeasible if it exceeds the available qubit count or requires deep circuits prone to decoherence. For instance, amplitude encoding is efficient in terms of qubit usage but often incurs significant circuit depth, rendering it less practical for near-term devices.

**Error rates** constitute a second major concern. Unlike classical operations, quantum gates are inherently probabilistic and are affected by various sources od noise, including environmental interactions, gate imperfections, and readout errors. Encoding schemes that amplify sensitivity to such noise, or that fail to incorporate error-aware design principles, may degrade the fidelity of the encoded quantum state. In this context, robustness to noise becomes a critical metric for evaluating the viability of an encoding method.

The **structure and complexity** of the data also play a central role in selecting an appropriate encoding strategy. Real-world data span a wide spectrum—from simple binary features to complex, high-dimensional objects such as images, audio signals, or graph-structured data. Some encoding schemes, such as basis encoding, are well-suited for discrete binary inputs but may struggle to efficiently represent continuous or structured data. Choosing a method that respects the geometry and statistics of the data is essential for preserving information during the encoding process.

Another key factor is the **computational overhead** associated with encoding—if the time and classical preprocessing required to prepare quantum states exceed the savings gained from quantum computation, *the overall advantage of the quantum approach is undermined.* This is particularly relevant for hybrid quantum–classical algorithms, where data must be loaded and encoded at each training iteration or inference step. Efficient, scalable encoding routines are therefore indispensable for achieving real-world speedups.

Finally, **reversibility** is a unique and often underappreciated constraint in quantum processing. Since quantum operations are unitary and thus reversible, encoding procedures should ideally preserve this property. Reversibility becomes especially important when the goal is not merely to process data but also to interpret the quantum output in a classical context. Ensuring that the encoding can be inverted allows for downstream tasks such as data reconstruction, debugging, or interpretability in hybrid workflows.

In summary, quantum encoding is not a purely theoretical exercise, but a design problem deeply intertwined with hardware capabilities, noise resilience, data structure, and computational efficiency. Developing encoding strategies that balance these considerations will be crucial to realizing the practical promise of quantum machine learning and quantum-enhanced data processing. As the field progresses beyond the NISQ era, these pragmatic constraints will continue to inform the co-design of quantum algorithms and hardware.

## 3.6 Summary

This chapter explored the foundational role of information encoding in quantum machine learning, emphasizing its theoretical underpinnings and practical implications. Encoding classical or quantum data into quantum states is a critical first step in any quantum learning pipeline, as it determines the hypothesis class accessible to quantum models and directly influences the trainability, expressivity, and computational efficiency of the resulting algorithms.

We began by reviewing classical approaches to feature representation and kernel-based learning, and then formalized their quantum counterparts through parameterized unitaries acting on multi-qubit systems. Several common encoding strategies were introduced, including basis, amplitude, angle, and Hamiltonian based encodings, each with distinct trade-offs in terms of qubit efficiency, circuit depth, and compatibility with current quantum hardware.

We then examined the concept of quantum autoencoders, which learn to compress quantum states into smaller subspaces while retaining relevant information. These models demonstrate how encoding can be learned, rather than manually prescribed, opening the door to task-specific or data-driven quantum representations. Their applications span quantum state compression, error mitigation, and generative modeling.

The chapter also introduced the notion of quantum data itself, not merely as encoded classical inputs but as genuine quantum states that arise from physical processes, simulations, or quantum circuits. This distinction brings forth new challenges in accessing, labeling, and learning from quantum data under realistic constraints such as the no-cloning theorem and measurement destructiveness.

Finally, we discussed a range of practical considerations in designing encoding schemes These include hardware limitations (qubit count, noise), the complexity of classical data, the overhead of encoding circuits, and the need for reversible operations. As the field moves toward more complex and scalable quantum systems, encoding will remain a central concern in co-designing quantum algorithms that are both expressive and implementable on near-term devices.

In summary, information encoding bridges the gap between the mathematical formalism of quantum theory and the practical demands of quantum computing. A nuanced understanding of encoding strategies is essential for developing effective and efficient quantum machine learning applications.

## References

[1] Coppersmith D 2002 An approximate Fourier transform useful in quantum factoring arXiv: quant-ph/0201067

[2] Schuld M and Petruccione F 2018 *Supervised Learning with Quantum Computers* (Cham: Springer International)

[3] Schuld M, Bocharov A, Svore K M and Wiebe N 2020 Circuit-centric quantum classifiers *Phys. Rev.* A **101** 032308

[4] Pérez-Salinas A, Cervera-Lierta A, Gil-Fuster E and Latorre J I 2020 Data re-uploading for a universal quantum classifier *Quantum* **4** 226

[5] Grover L and Rudolph T 2002 Creating superpositions that correspond to efficiently integrable probability distributions arXiv: quant-ph/0208112

[6] McClean J R, Boixo S, Smelyanskiy V N, Babbush R and Neven H 2018 Barren plateaus in quantum neural network training landscapes *Nat. Commun.* **9** 4812

[7] Schuld M and Killoran N 2019 Quantum machine learning in feature Hilbert spaces *Phys. Rev. Lett.* **122** 040504

[8] Rebentrost P, Mohseni M and Lloyd S 2014 Quantum support vector machine for big data classification *Phys. Rev. Lett.* **113** 130503

[9] Rumelhart D E, Hinton G E and Williams R J 1986 Learning internal representations by error propagation *Parallel Distributed Processing: Explorations in the Microstructure of Cognition: Foundations* (Cambridge, MA: MIT Press) pp 318–62

[10] Bank D, Koenigstein N and Giryes R 2023 Autoencoders *Machine Learning for Data Science Handbook: Data Mining and Knowledge Discovery Handbook* (Cham: Springer International) pp 353–74

[11] Romero J, Olson J P and Aspuru-Guzik A 2017 Quantum autoencoders for efficient compression of quantum data *Quantum Sci. Technol.* **2** 045001

[12] Khoshaman A, Vinci W, Denis B, Andriyash E, Sadeghi H and Amin M H 2018 Quantum variational autoencoder *Quantum Sci. Technol.* **4** 014001

[13] Wan K H, Dahlsten O, Kristjánsson H, Gardner R and Kim M S 2017 Quantum generalisation of feedforward neural networks *npj Quantum Inf.* **3** 36

[14] Mangini S, Marruzzo A, Piantanida M, Gerace D, Bajoni D and Macchiavello C 2022 Quantum neural network autoencoder and classifier applied to an industrial case study *Quant. Mach. Intell.* **4** 13

[15] Ngairangbam V S, Spannowsky M and Takeuchi M 2022 Anomaly detection in high-energy physics using a quantum autoencoder *Phys. Rev.* D **105** 095004

[16] Zhang X-M, Kong W, Farooq M U, Yung M-H, Guo G and Wang X 2021 Generic detection-based error mitigation using quantum autoencoders *Phys. Rev.* A **103** L040403

[17] Pazem J and Ansari M H 2023 Error mitigation of entangled states using brainbox quantum autoencoders *Sci. Rep.* **15** 2257

[18] Sentís G, Monràs A, Muñoz Tapia R, Calsamiglia J and Bagan E 2019 Unsupervised classification of quantum data *Phys. Rev.* X **9** 041029

[19] Wang Y and Cao L 2025 Quantum phase transition detection via quantum support vector machine *Quantum Sci. Technol.* **10** 015043

[20] Xia Y, Li W, Zhuang Q and Zhang Z 2021 Quantum-enhanced data classification with a variational entangled sensor network *Phys. Rev.* X **11** 021047

[21] Kardashin A, Balkybek Y, Palyulin V V and Antipin K 2025 Predicting properties of quantum systems by regression on a quantum computer *Phys. Rev. Res.* **7** 013201

[22] Zoufal C 2021 Generative quantum machine learning arXiv: 2111.12738

[23] O'Donnell R and Wright J 2015 Efficient quantum tomography arXiv: 1508.01907

[24] O'Donnell R and Wright J 2017 Efficient quantum tomography II *49th Annual ACM SIGACT Symp. on Theory of Computing* pp 899–912

[25] García-Pérez G, Rossi M A C, Sokolov B, Tacchino F, Barkoutsos P K, Mazzola G, Tavernelli I and Maniscalco S 2021 Learning to measure: adaptive informationally complete generalized measurements for quantum algorithms *PRX Quantum* **2** 040342

**IOP** Publishing

Quantum Machine Learning
Concepts and possibilities
**Andrea Delgado and Kathleen E Hamilton**

# Chapter 4

## Quantum computing for inference

Supervised learning is a central paradigm in classical machine learning, wherein a model learns a mapping from inputs to outputs using labeled data. From image classification to spam detection, supervised learning algorithms are widely used to extract patterns from data and make reliable predictions on unseen examples. At the heart of these methods lie models, often linear or nonlinear, that learn decision boundaries based on the training data, and tools such as feature maps and kernels that enhance the model's ability to capture complex relationships in high-dimensional spaces.

With the advent of quantum computing, researchers have begun to explore how quantum mechanical principles can be harnessed to enhance the performance of inference algorithms. Quantum computers, by operating in Hilbert spaces of exponentially large dimension, offer a new substrate for representing and processing information. In particular, quantum machine learning introduces a novel approach to supervised learning by replacing classical components, such as feature maps or inner products, with their quantum analogs. These include **quantum feature maps**, which encode classical data into quantum states, and **quantum kernels**, which evaluate the similarity of data points via quantum circuits.

The key insights is that certain quantum feature maps can implicitly embed data into extremely high-dimensional spaces, where learning tasks may become easier, and where classical computers would struggle to compute the corresponding inner products. This opens the possibility of **quantum-enhanced inference**, where quantum circuits help extract patterns from data in a way that is more expressive or efficient than classical models, at least in principle.

In this chapter, we explore the role of quantum computing in supervised learning from the perspective of **inference**. We begin with a brief review of classical inference, feature maps, and kernel methods. We then introduce the notion of quantum feature maps and show how they induce quantum kernels that can be used within kernel-based learning algorithms such as support vector machines (SVMs). We also discuss **parameterized quantum circuits (PQCs)** that act as linear models in Hilbert space and can be trained to perform classification tasks.

doi:10.1088/978-0-7503-4952-9ch4  

Along the way, we provide insight into both the theoretical motivations and practical implementations of quantum inference models, including discussion of current empirical benchmarks, noise robustness, and the open question of quantum advantage. By the end of this chapter, the reader will be equipped with a foundational understanding of how quantum devices can be used for supervised learning, the potential benefits they may offer, and the challenges that must be overcome to realize practical quantum inference.

## 4.1 Classical foundations of inference

To understand how quantum computing can aid inference tasks, we must first revisit the classical framework of supervised learning and kernel-based models. These foundations not only guide the design of quantum analogs but serve as benchmarks to evaluate the performance and expressivity of quantum approaches.

### 4.1.1 Supervised learning basics

Supervised learning refers to the process of learning a function that maps inputs to outputs based on label examples. Formally, we are giving a training dataset

$$\mathscr{D} = \left\{(x_i, y_i)\right\}_{i=1}^{n}, \tag{4.1}$$

where each $x_i \in \mathbb{R}^d$ is a feature vector and $y_i \in \mathscr{Y}$ is the corresponding label. For classification problems, the label set is finite, typically $\mathscr{Y} = \{1, \ldots, n\}$, whereas for regression, $\mathscr{Y} = \mathbb{R}$.

The learning algorithm aims to find hypothesis $h: \mathbb{R}^d \to \mathscr{Y}$ within a chosen hypothesis class $\mathscr{H}$, such that the model generalizes well to new, unseen data. The generalization ability is typically assessed via a loss function $\mathscr{L}(h(x), y)$, such as the 0–1 loss for classification or the squared loss for regression. The empirical risk (or training error) is defined as

$$\hat{R}(h) = \frac{1}{n}\sum\nolimits_{i=1}^{n} \mathscr{L}(h(x_i), y_i), \tag{4.2}$$

and the goal is to minimize the expected risk (test error) over the data distribution.

A fundamental challenge in classical supervised learning is balancing *bias* and *variance*: models that are too simple may underfit the data (high bias), while overly complex models may overfit (high variance). Regularization, cross-validation, and model selection techniques are typically used to navigate this trade-off.

### 4.1.2 Feature maps and kernels in classical machine learning

Many learning tasks benefit from transforming the data into a new space where the structure becomes more amenable to linear separation. This transformation is performed by a **feature map**:

$$\phi: \mathbb{R}^d \to \mathscr{F}, \tag{4.3}$$

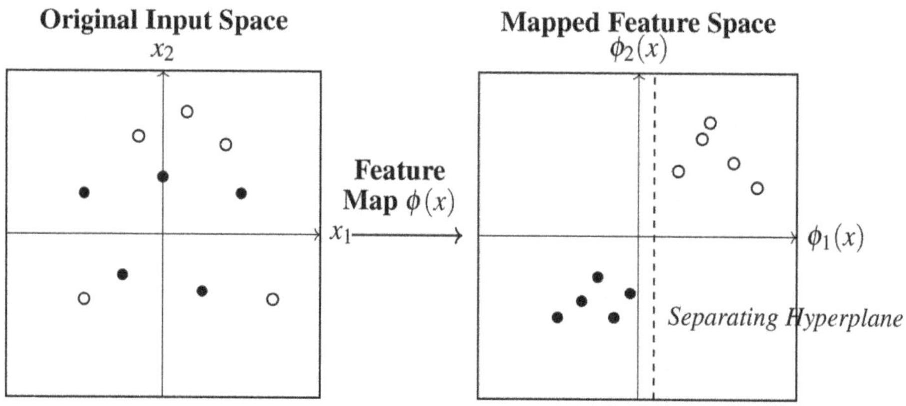


**Figure 4.1.** Illustration of the kernel trick: a nonlinear transformation $\phi(x)$ maps non-separable data in input space to a higher-dimensional feature space where they become linearly separable.

where $\mathscr{F}$ is a (possibly high- or infinite-dimensional) feature space. A simple linear model in feature space is then given by

$$h(x) = \langle w, \phi(x) \rangle + b, \tag{4.4}$$

where $w \in \mathscr{F}$ and $b \in \mathbb{R}$. If $\phi$ is appropriately chosen, even a linear model can capture nonlinear patterns in the input space (see figure 4.1).

### 4.1.3 Kernel methods

Instead of explicitly computing $\phi(x)$, *kernel methods* evaluate inner products in the feature space using a *kernel function*

$$k(x, x') = \langle \phi(x), \phi(x') \rangle. \tag{4.5}$$

This trick—known as the *kernel trick*—allows efficient computation in high-dimensional space. A popular example is the radial basis function (RBF) kernel,

$$k(x, x') = \exp\left(-\frac{|x - x'|^2}{2\sigma^2}\right), \tag{4.6}$$

which corresponds to an infinite-dimensional feature space.

Kernel methods such as the SVM make use of kernel functions to construct classifiers with strong generalization properties. The dual formulation of the SVM optimization problem depends only on kernel evaluations, making it possible to learn complex decision boundaries without ever explicitly constructing the feature vectors.

### 4.1.4 Theoretical guarantees

The use of kernel methods is grounded in the *representer theorem*, which ensures that solutions to a wide class of regularized empirical risk minimization problems lie in the span of the kernel evaluations on the training set. Furthermore, *Mercer's theorem* guarantees that any continuous, symmetric, positive-definite kernel corresponds to an inner product in some Hilbert space.

**The kernel trick**

Given a feature map $\phi\colon \mathbb{R}^d \to \mathcal{F}$, the kernel trick allows us to compute inner products in $\mathcal{F}$ without explicitly constructing $\phi(x)$. That is,

$$k(x, x') = \langle \phi(x), \phi(x') \rangle \tag{4.7}$$

can be computed directly in the input space. This leads to powerful algorithms that operate in high-dimensional or even infinite-dimensional feature spaces with computational cost depending only on the size of the dataset.

These classical techniques motivate quantum counterparts. By replacing classical feature maps and inner products with quantum states and overlap amplitudes, one can define **quantum kernels** and **quantum-enhanced inference models**. The next sections will explore how quantum computing implements these ideas, with the potential for exponential expressivity and computational advantages.

## 4.2 Quantum feature maps

One of the most promising avenues for leveraging quantum computing in supervised learning is the use of **quantum feature maps**. These maps embed classical data into high-dimensional quantum states, opening the possibility of using quantum hardware to implicitly access large Hilbert spaces that are intractable to simulate classically. The inner products between such quantum states serve as the foundation for **quantum kernel methods** and other quantum-enhanced inference algorithms.

### 4.2.1 Quantum states as features

Let us recall the concept of a classical feature map $\phi\colon \mathbb{R}^d \to \mathcal{F}$, where $\mathcal{F}$ is a high-dimensional feature space. In the quantum setting, the role of $\phi(x)$ is played by a quantum state $|\phi(x)\rangle$, prepared by a quantum circuit that encodes the classical data $x$.

Formally, a *quantum feature map* is a mapping

$$x \mapsto |\phi(x)\rangle \in \mathcal{H}, \tag{4.8}$$

where $\mathcal{H}$ is the Hilbert space associated with an $n$-qubit quantum system, i.e. $\mathcal{H} \simeq (\mathbb{C}^2)^{\otimes n}$. The quantum circuit that prepares $|\phi(x)\rangle$ is often denoted by $U_\phi(x)$, such that

$$|\phi(x)\rangle = U_\phi(x)|0\rangle^{\otimes n}. \tag{4.9}$$

The encoding circuit $U_\phi(x)$ may include data-dependent gates such as $R_z(x_j)$, $R_y(x_j)$, or more expressive encodings involving Hamiltonian evolution or entangling operations. The choice of encoding affects the geometry of the quantum feature space and ultimately the performance of quantum inference algorithms.

### 4.2.2 Data encodings for quantum feature maps

Several encoding strategies were introduced in section 3.2. Here, we briefly recall their relevance in the context of feature maps:

- **Basis encoding:** Encodes binary strings directly into the computational basis. For example, the string $x = 101$ is mapped to $|x\rangle = |1\rangle \otimes |0\rangle \otimes |1\rangle$. This is efficient but limited in expressivity.
- **Amplitude encoding:** Represents a normalized vector $x \in \mathbb{R}^d$ as a quantum state:

$$x \mapsto |x\rangle = \sum_{j=1}^{d} x_j |j\rangle, \tag{4.10}$$

  where $\sum_j |x_j|^2 = 1$. Amplitude encoding is exponentially compact, although state preparation can be costly.
- **Angle encoding:** Each feature $x_j$ is mapped to a rotation angle on a qubit:

$$x_j \mapsto R_y(x_j)|0\rangle = \cos(x_j/2)|0\rangle + \sin(x_j/2)|1\rangle. \tag{4.11}$$

- **Hamiltonian encoding:** Applies a data-dependent unitary $U(x) = e^{-iH(x)}$, where $H(x)$ is a Hamiltonian encoding the data. This strategy is commonly used in quantum kernel methods.

Each encoding scheme offers trade-offs in terms of circuit depth, representational power, and robustness to noise. In practice, angle and Hamiltonian encodings are popular choices for constructing expressive and hardware-feasible quantum feature maps.

### 4.2.3 Properties of quantum feature maps

Quantum feature maps are designed to exploit the geometric structure of Hilbert space to encode data in ways that are:

- **Expressive:** Quantum feature maps may encode nonlinear relationships in a highly entangled quantum state space, enabling more complex decision boundaries.
- **Hard to simulate classically :** If the quantum inner product $|\langle\phi(x)|\phi(x')\rangle|^2$ is classically hard to compute (under plausible complexity assumptions), then the associated kernel function may be inaccessible to classical methods.
- **Implicitly high-dimensional:** Due to the exponential size of the Hilbert space, quantum states can implicitly represent feature vectors in space with dimension $2^n$, where $n$ is the number of qubits. Unlike classical machine learning, the inner product can be computed via a physical quantum process.

**Table 4.1.** Comparison of classical and quantum feature maps.

| Property | Classical feature map $\phi(x)$ | Quantum feature map $\|\phi(x)\rangle$ |
|---|---|---|
| Domain | $\mathbb{R}^d \to \mathscr{F}$ | $\mathbb{R}^d \to \mathscr{H} = (\mathbb{C}^2)^{\otimes n}$ |
| Representation | Explicit vector | Quantum state prepared by a unitary circuit |
| Dimensionality | Often infinite (e.g. RBF kernels) | Exponential in number of qubits |
| Inner product | $\langle \phi(x), \phi(x') \rangle$ | $\|\langle \phi(x) \| \phi(x') \rangle\|^2$ |
| Evaluation cost | Polynomial or exponential in dim $\mathscr{F}$ | Depends on quantum circuit depth and measurement shots |
| Potential benefit | Improved separability | Compact and expressive embeddings in Hilbert space |

These properties make quantum feature maps central to many quantum machine learning algorithms. In table 4.1, we present a comparison of classical and quantum feature maps to highlight similarities and potential advantages of each domain. In the next section, we will see how quantum feature maps naturally induce **quantum kernels**, which form the backbone of quantum kernel methods for supervised learning.

## 4.3 Quantum kernels and kernel methods

The concept of a kernel is central to many classical inference algorithms, particularly in SVMs and Gaussian processes. When data are encoded into quantum states via a quantum feature map $x \mapsto |\phi(x)\rangle$, the inner product between these states defines a *quantum kernel*. Quantum kernels offer a natural extension of classical kernel methods into the quantum regime, with the potential to represent highly expressive similarity functions that are difficult to compute classically.

### 4.3.1 Quantum kernel definition

Given a quantum feature map $|\phi(x)\rangle$, the associated quantum kernel is defined as

$$\kappa(x, x') = |\langle \phi(x) | \phi(x') \rangle|^2. \tag{4.12}$$

This expression quantifies the fidelity between two quantum states prepared by circuits $U_\phi(x)$ and $U_\phi(x')$. The higher the fidelity, the more similar the inputs $x$ and $x'$ are considered to be, according to the geometry of the quantum Hilbert space induced by the encoding.

An alternative (but less common) definition used the real part of the overlap,

$$\kappa(x, x') = \mathrm{Re}(\langle \phi(x) | \phi(x') \rangle), \tag{4.13}$$

although this form is less commonly used in practice due to measurement constraints.

### 4.3.2 Evaluating quantum kernels

In practice, the quantum kernel $\kappa(x, x')$ is evaluated using a quantum circuit. Two common strategies are:

- **The swap test**: A circuit that estimates the overlap $|\langle\phi(x)|\phi(x')\rangle|^2$ by using an ancilla qubit to interfere two quantum states. This requires controlled-swap gates and multiple copies of the states.
- **Direct fidelity estimation**: If both states $|\phi(x)\rangle$ and $|\phi(x')\rangle$ are prepared on the same device, the fidelity can be estimated by preparing the state $U_\phi(x)^\dagger U_\phi(x')|0\rangle$ and measuring the probability of obtaining the all-zero state.

After computing all pairwise kernel values over a training set, one obtains the kernel matrix $K$, where $K_{ij} = \kappa(x_i, x_j)$. This matrix is then used as input to a classical kernel method such as an SVM, yielding a hybrid quantum–classical learning pipeline.

### 4.3.3 Training with quantum kernels

Once the kernel matrix is computed, the training procedure follows that of a standard classical kernel method. For instance, in an SVM classifier, the optimization problem involves solving a quadratic program,

$$\min_\alpha \frac{1}{2}\sum\nolimits_{i,j}\alpha_i\alpha_j y_i y_j \kappa(x_i, x_j) - \sum\nolimits_i \alpha_i, \tag{4.14}$$

subject to the constraints $0 \leqslant \alpha_i \leqslant C$ and $\sum_i \alpha_i y_i = 0$, where $\alpha_i$ are dual variables and $C$ is a regularization parameter.

Importantly, this training is performed on a classical computer. The quantum device is only used to compute the kernel matrix $K$, which captures the geometry of the data in Hilbert space.

### 4.3.4 Generalization and expressivity

A key question in quantum machine learning is whether quantum kernels offer any statistical or computational advantage over classical kernels. Several recent studies have analysed this question through the lens of:

- **Kernel alignment**: Measures how well the quantum kernel matrix aligns with the ideal kernel that separates the data. Poor alignment can result in underperformance compared to classical kernels.
- **Inductive bias and reproducing kernel Hilbert space (RKHS)**: Every positive-definite kernel defines an RKHS. The hypothesis class of quantum kernel methods is constrained by the RKHS associated with $\kappa(x, x')$, which determines the types of functions that can be learned.
- **Hardness of classical approximation**: If $\kappa(x, x')$ is difficult to approximate classically (e.g. related to problems such as random circuit sampling or instantaneous quantum polynomial-time (IQP) models), the quantum kernel could provide a separation from classical methods.

**Pros and cons of quantum kernels**
**Pros**:
- Can implicitly embed data into exponentially large Hilbert spaces.
- May be hard to simulate classically under complexity-theoretic assumptions.
- Naturally compatible with existing classical kernel learning frameworks.

**Cons**:
- Kernel estimation requires many quantum circuit evaluations (shots).
- Sensitivity to noise can degrade fidelity estimates.
- Poor kernel alignment can lead to suboptimal performance.

In the next section, we move beyond fixed-feature kernel models and explore *linear quantum models*, in which the hypothesis class is encoded directly into a PQC whose parameters are learned during training.

## 4.4 Linear quantum models

Quantum kernel methods rely on fixed quantum feature maps and classical optimization. In contrast, *linear quantum models* introduce trainable parameters directly into quantum circuits. These models represent quantum analogs of linear classifiers, such as logistic regression or perceptrons, but operate in high-dimensional Hilbert space where the feature space is defined by the structure of a quantum circuit.

The most common approach involves PQcs, also known as variational quantum circuits, where some gates are equipped with tunable angles. These parameters are optimized using classical routines, making the overall learning architecture a hybrid quantum–classical model.

### 4.4.1 Parameterized quantum circuits for classification

A parameterized quantum circuit is a quantum circuit with gates that depend on a vector of parameters $\boldsymbol{\theta}$. In the context of classification, a typical circuit structure includes:

- A quantum feature map $U_\phi(x)$, which encodes the classical input $x$ into a quantum state.
- A trainable unitary $U(\boldsymbol{\theta})$, applied after the encoding, representing the model's hypothesis.
- A measurement in the computational basis to extract the output.

This circuit prepares a state

$$|\psi(x, \boldsymbol{\theta})\rangle = U(\boldsymbol{\theta})U_\phi(x)|0\rangle^{\otimes n}, \tag{4.15}$$

and the predicted label is obtained by measuring one or more qubits and post-processing the result. For example, in binary classification, one might define

$$f(x; \boldsymbol{\theta}) = \langle\psi(x, \boldsymbol{\theta})|M|\psi(x, \boldsymbol{\theta})\rangle, \tag{4.16}$$

where $M = Z \otimes \mathbb{I} \otimes \cdots \otimes \mathbb{I}$ is a Pauli operator acting on the first qubit. A decision rule is then applied, e.g. classify $x$ as class 1 if $f(x; \boldsymbol{\theta}) > 0$ and class 0 otherwise.

### 4.4.2 Training strategies

The model parameters $\boldsymbol{\theta}$ are optimized by minimizing a cost function $\mathcal{L}(\boldsymbol{\theta})$, such as cross-entropy or mean squared error, over a training dataset. Optimization is typically performed using a classical algorithm such as gradient descent or one of its variants.

There are two main approaches for optimization:

- **Gradient-based optimization**: Uses analytical gradients computed via the *parameter-shift rule*, which allows exact derivative estimates from additional circuit evaluations.
- **Gradient-free optimization**: Includes methods such as COBYLA, Nelder–Mead, or genetic algorithms. These are often more robust to noise but may converge slowly.

The cost landscape can exhibit *barren plateaus*—regions where the gradient is vanishingly small—particularly in deep or highly expressive circuits. This makes the choice of circuit architecture and initialization critical for successful training.

### 4.4.3 Example: variational quantum classifier

A common implementation of a linear quantum model is the *variational quantum classifier* (VQC). It typically follows the architecture:

1. Apply angle encoding $U_\phi(x)$ to map input $x$ to a quantum state.
2. Apply a parametrized circuit $U(\boldsymbol{\theta})$ consisting of layers of single-qubit rotations and entangling gates.
3. Measure a subset of qubits in the $Z$-basis.
4. Use the measurement outcomes (or their expectation values) as the model's output.

The VQC is expressive and trainable and can be adapted to binary or multiclass classification tasks. However, care must be taken to avoid overparameterization or insufficient entanglement, both of which affect generalization.

**Remarks in variational circuits**

- The expressivity of a PQC depends on its depth, the choice of entangling gates, and the connectivity between qubits.
- Shallow circuits may underfit the data, while deep circuits are prone to barren plateaus.
- Measurement strategies affect both the efficiency of training and the interpretability of results.

Linear quantum models offer a trainable alternative to fixed-feature quantum kernels. While kernel methods rely on evaluating quantum overlaps, variational models attempt to learn the mapping directly in Hilbert space by tuning circuit parameters. In the next section, we compare the performance of quantum inference models—both kernel-based and variational—on benchmark tasks and discuss their potential.

## 4.5 Performance and benchmarks

Understanding the practical utility of quantum inference models requires a critical evaluation of their performance across multiple dimensions. In this section, we examine the theoretical motivations for using quantum models in supervised learning, summarize recent benchmark studies, and discuss how hardware noise and limited qubit counts influence model performance on near-term devices.

### 4.5.1 Theoretical speedups and limitations

The appeal of quantum machine learning stems from the potential for computational speedups and richer representational power. In the context of supervised learning, this potential is reflected in three key ideas:

- **Implicit exponential feature spaces:** Quantum circuits can map classical data into quantum states that inhabit exponentially large Hilbert spaces. This enables highly expressive hypothesis classes without the need to explicitly construct large feature vectors [1–4].
- **Hard-to-simulate kernels:** Some quantum kernels are conjectured to be classically intractable to compute under standard complexity-theoretic assumptions. This inaccessibility may allow quantum models to generalize better on problems where classical models are fundamentally limited [5, 6].
- **Linear speedups in optimization:** For variational quantum models, gradients can often be estimated efficiently using the parameter-shift rule [7, 8], leading to training times that scale linearly with the number of parameters under favorable conditions.

That said, many of these advantages assume ideal, fault-tolerant quantum hardware. On noisy intermediate-scale quantum (NISQ) devices, empirical benchmarking is crucial to assess what can be achieved in practice.

### 4.5.2 Empirical studies and benchmarks

Recent benchmarking efforts have compared quantum inference models to classical baselines across both synthetic and real datasets. Key observations include:

- On simple low-dimensional datasets (e.g. XOR [9, 10], concentric circles [11], parity functions [12]), quantum kernel methods and variational quantum classifiers can match or slightly outperform classical models when circuits are designed carefully [13].
- On larger or real-world datasets (e.g. MNIST digits [14]), classical models such as SVMs and neural networks tend to outperform quantum models due

to the limited number of qubits, circuit depth constraints, and hardware noise [15].
- Hybrid workflows [16, 17] that combine quantum subroutines (e.g. kernel estimation) with classical optimization (e.g. SVM training) show more robustness and scalability when regularized properly [18].

Evaluation metrics in benchmark studies typically include:
- **Accuracy and generalization** on held-out sets.
- **Data efficiency**, measuring performance as a function of training set size.
- **Robustness to noise**, assessing how quantum circuits behave under hardware errors.
- **Training efficiency**, evaluating the number of circuit evaluations required.

### 4.5.3 Robustness and noise considerations

Quantum inference on NISQ devices faces several physical and algorithmic challenges:
- **Decoherence and gate errors** reduce the effective circuit depth, limiting the complexity of models that can be trained in practice.
- **Shot noise** introduces statistical fluctuations due to probabilistic measurements. Accurate estimation of expectation values often requires thousands of repeated measurements, which increases runtime.
- **Error mitigation techniques** such as zero-noise extrapolation and readout correction can help, but add computational overhead and may not scale well.

To address these limitations, researchers often simulate noise during training or pre-train parameters classically before deploying circuits on hardware. Some methods even optimize training by modeling noise in the cost function explicitly.

In summary, there is currently no conclusive experimental evidence of quantum advantage for supervised learning on real-world datasets using available quantum hardware. Some theoretical constructions suggest potential quantum–classical separation in contrived or worst-case distributions, usually via hard-to-simulate quantum kernels. In practice, quantum advantage is more likely to emerge in hybrid workflows co-designed with application-specific constraints and error mitigation strategies.

## 4.6 Case studies and applications

While theoretical and benchmark studies provide insight into the capabilities and limitations of quantum inference models, their practical relevance is well understood through domain-specific applications. This section presents case studies that illustrate how quantum kernel methods and variational quantum classifiers have been applied to real-world data from a variety of scientific and industrial domains. These examples help identify both opportunities for near-term deployment and challenges that remain for achieving quantum advantage.

### 4.6.1 Drug discovery and molecular property prediction

Quantum inference methods have found early applications in the pharmaceutical industry, particularly for predicting molecular properties such as solubility, toxicity, and bioactivity. These tasks are typically posed as binary or multiclass classification problems, where the input is a molecular descriptor or graph-based encoding, and the output is a label indicating a chemical or biological property.

Quantum kernel methods have been applied to the QM7 and QM9 datasets (quantum chemistry benchmarks), where quantum circuits encode molecular fingerprints and compute similarity via fidelity-based kernels. Studies have shown comparable performance to classical kernels when carefully regularized and noise-mitigated [19].

Variational quantum classifiers trained on binary classification tasks involving molecular datasets (e.g. predicting HIV inhibition) using angle-encoded classical features. These models demonstrate feasibility on NISQ hardware but remain limited by circuit depth and expressivity [20, 21].

Despite the current performance gap with classical deep learning models, these studies demonstrate the potential for quantum models to serve as subroutines in larger quantum chemistry workflows, in particular where quantum structure can be directly encoded.

### 4.6.2 High energy physics

In experimental high-energy physics (HEP), machine learning is widely used for classifying particle signatures in collider experiments. Quantum inference models have been explored as alternatives or supplements to classical classifiers in scenarios such as *jet tagging* in events from the Large Hadron Collider (LHC), where quantum models classify jets as originating from particles based on high-dimensional event features [22].

Quantum kernel methods [23, 24] and variational classifiers trained on simulated HEP datasets have shown comparable performance to shallow classical models, in particular when the input dimensionality is reduced via preprocessing or principal component analysis (PCA) [25].

HEP applications are particularly attractive because they involve high-dimensional, highly structured data, offering an opportunity to leverage the expressivity of quantum models. However, achieving state-of-the-art performance remains challenging given the maturity of the classical pipeline in this field.

### 4.6.3 Finance

In the financial sector, quantum kernel methods have been proposed for detecting anomalies in transaction networks by computing fidelity-based similarity scores between customer profiles [26].

These applications benefit from the ability of quantum models to capture nonlinear correlations in noisy, structured datasets. However, deployment remains at the proof-of-concept stage due to stringent performance requirements and regulatory considerations.

**Common traits among applications**

- Tasks are often low-data or high-dimensional—favorable settings for kernel-based methods.
- Preprocessing (e.g. PCA, downsampling, feature selection) is critical to adapting data to current quantum hardware.
- Quantum advantage has not yet been demonstrated on real-world datasets, but hybrid methods show promise.
- Most implementations remain limited to binary classification due to circuit and qubit constraints.

These case studies illustrate the diversity of application domains in which quantum inference models are being tested. Although limitations persist, in particular with respect to hardware capabilities and model expressivity, these real-world experiments are essential for guiding co-design strategies between quantum algorithms, data encodings, and physical devices.

## 4.7 Open challenges and future directions

Quantum inference is a rapidly evolving field situated at the intersection of quantum computing and machine learning. While early results have demonstrated feasibility and potential, there remain significant theoretical, algorithmic, and hardware-related challenges that must be addressed before quantum models can consistently outperform classical ones in practical tasks. In this section, we outline several key open problems and emerging research directions.

### 4.7.1 Scalability and resource constraints

The most immediate obstacle to deploying quantum inference models at scale is the limited size and fidelity of quantum hardware. Current devices are constrained by:

- **Qubit count:** Many practical datasets require input features that far exceed the number of available qubits. Dimensionality reduction and feature compression are necessary, but may limit the model's expressivity.
- **Circuit depth:** Deep circuits are needed to realize expressive quantum feature maps and variational ansatzes, but are often infeasible due to noise and decoherence.
- **Sampling cost:** Estimating expectation values requires many repetitions (shots), in particular for small signal-to-noise ratios. This increases training time and reduces model efficiency.

Research into more hardware-efficient architectures (e.g. shallow circuits, tensor networks, or error-mitigated inference schemes) is essential to overcoming these constraints.

### 4.7.2 Design of quantum feature maps

The choice of quantum feature map has a critical impact on the performance of both kernel-based and variational models. However, there is no universal method for constructing feature maps that guarantee good generalization. Open questions include:

- How do different encodings affect the geometry of the Hilbert space and the learnability of the target function?
- Can one systematically design quantum feature maps with provable advantages over classical counterparts?
- What classes of problems benefit most from quantum feature spaces?

Understanding the expressivity and inductive biases of quantum feature maps remains a major theoretical and empirical challenge.

### 4.7.3 Noise robustness and error mitigation

Quantum circuits used for inference are inherently noisy, and even shallow circuits can suffer from decoherence, gate infidelity, and readout errors. While some methods (e.g. zero-noise extrapolation, randomized compiling) can reduce these effects, they add computational overhead and are not yet standard across platforms. Key directions include:

- Developing noise-aware training algorithms that are robust to stochastic loss landscapes.
- Formalizing the impact of noise on generalization bounds and learning guarantees.
- Exploring hybrid or variational error mitigation methods tailored for inference tasks.

### 4.7.4 Learning theory and generalization bounds

Despite recent progress in benchmarking and empirical evaluation, there is a lack of rigorous theory of quantum models. Open theoretical questions include:

- What are the sample complexity and generalization bounds on quantum classifiers?
- Can quantum models provably reduce the number of training examples needed to achieve a given accuracy?
- Under what conditions can quantum models be said to learn more efficiently than classical ones?

These questions are fundamental to understanding when and why quantum models might outperform their classical counterparts, particularly in regimes of limited data or high-dimensional structure.

### 4.7.5 Hybrid quantum–classical workflows

Many promising results have been obtained by combining quantum and classical components into hybrid learning pipelines. However, best practices for co-design are still being established. Research is needed on the efficient division of labor between quantum and classical modules, integration with existing classical machine learning pipelines, including preprocessing and post-processing steps, and meta-learning and automated architecture search for hybrid models. These efforts could guide the development of quantum inference systems that work alongside classical tools in realistic machine learning settings.

## 4.8 Summary

In this chapter, we explored how quantum computing can be applied to the task of supervised inference, where the goal is to learn a model from labeled data that generalize to unseen inputs. By drawing analogies with classical machine learning, we introduced key quantum components—such as quantum feature maps, quantum kernels, and variational quantum circuits—and examined how they can be used to construct powerful inference models.

We began by reviewing the classical foundations of inference, emphasizing the role of feature maps and kernel methods in constructing models that capture complex patterns. These ideas naturally motivated the quantum analogs: quantum feature maps that encode classical data into quantum states, and quantum kernels that measure the similarity between these states via inner products in Hilbert space.

We then examined two major quantum approaches to supervised learning:

- **Quantum kernel methods**, where fixed quantum feature maps are used to define a kernel matrix for a classical learning algorithm such as an SVM.
- **Linear quantum models**, including variational quantum classifiers, where the decision function is implemented directly via trainable quantum circuits.

We discussed the relative strengths and weaknesses of each approach. Kernel methods are interpretable and integrate well with classical theory, while variational models offer greater flexibility and adaptability, in particular when data and task requirements are less well-understood.

Empirical benchmarks on both synthetic and real datasets indicate that while quantum inference models can match or slightly outperform classical baselines in specific cases, broad-scale advantage has not yet been demonstrated. Practical deployment remains limited by the depth, noise, and scalability of NISQ-era hardware.

Nevertheless, we reviewed several **application case studies**—including drug discovery, high-energy physics, and financial modeling—where quantum models are actively being explored and tested. These efforts highlight the potential for quantum inference as part of a hybrid computational strategy, particularly in data-scarce or high-dimensional settings.

Finally, we reflected on **open challenges** and future research directions, from better quantum feature map design to a deeper theoretical understanding of generalization and learning complexity in the quantum setting.

## References


[1] Schuld M and Petruccione F 2018 *Supervised Learning with Quantum Computers* (Cham: Springer International)

[2] Havlíček V, Córcoles A D, Temme K, Harrow A W, Kandala A, Chow J M and Gambetta J M 2019 Supervised learning with quantum-enhanced feature spaces *Nature* **567** 209–212

[3] Schuld M and Killoran N 2019 Quantum machine learning in feature Hilbert spaces *Phys. Rev. Lett.* **122** 040504

[4] Lloyd S, Schuld M, Ijaz A, Izaac J and Killoran N 2001 Quantum embeddings for machine learning arXiv: 2001.03622

[5] Gentinetta G, Thomsen A, Sutter D and Woerner S 2024 The complexity of quantum support vector machines *Quantum* **8** 1225

[6] Gil-Fuster E, Eisert J and Dunjko V 2024 On the expressivity of embedding quantum kernels *Mach. Learn.: Sci. Technol.* **5** 025003

[7] Wierichs D, Izaac J, Wang C and Yen-Yu Lin C 2022 General parameter-shift rules for quantum gradients *Quantum* **6** 677

[8] Schuld M, Bergholm V, Gogolin C, Izaac J and Killoran N 2019 Evaluating analytic gradients on quantum hardware *Phys. Rev.* A **99** 032331

[9] Rathi N and Kumar S 2025 Enhanced quantum ensemble classification algorithm with shallow circuit and parallelism mechanism *Concurr. Comput.: Pract. Exp.* **37** e70166

[10] Suzuki Y, Yano H, Gao Q, Uno S, Tanaka T, Akiyama M and Yamamoto N 2020 Analysis and synthesis of feature map for kernel-based quantum classifier *Quant. Mach. Intell.* **2** 9

[11] Schuld M, Fingerhuth M and Petruccione F 2017 Implementing a distance-based classifier with a quantum interference circuit *Europhys. Lett.* **119** 60002

[12] Sen P, Bhatia A S, Bhangu K S and Elbeltagi A 2022 Variational quantum classifiers through the lens of the hessian *PLoS One* **17** e0262346

[13] Hur T, Kim L and Park D K 2022 Quantum convolutional neural network for classical data classification *Quant. Mach. Intell.* **4** 3

[14] Butmaratthaya S, Buesamae N and Taetragool U 2023 MNIST quantum classification models implementation and benchmarking *AIP Conf. Proc.* **2906** 070002

[15] Wang H, Ding Y, Gu J, Lin Y, Pan D Z, Chong F T and Han S 2022 QuantumNAS: noise-adaptive search for robust quantum circuits *2022 IEEE Int. Symp. on High-Performance Computer Architecture (HPCA)* pp 692–708

[16] Sarkar S 2024 Quantum transfer learning for MNIST classification using a hybrid quantum-classical approach arXiv: 2408.03351

[17] Ranga D, Prajapat S, Akhtar Z, Kumar P and Vasilakos A V 2024 Hybrid quantum-classical neural networks for efficient MNIST binary image classification *Mathematics* **12** 3684

[18] Grant E, Benedetti M, Cao S, Hallam A, Lockhart J, Stojevic V, Green A G and Severini S 2018 Hierarchical quantum classifiers *npj Quantum Inf.* **4** 65

[19] Khan D, Heinen S and von Lilienfeld O A 2023 Kernel based quantum machine learning at record rate: many-body distribution functionals as compact representations *J. Chem. Phys.* **159** 034106

[20] Sathan D and Baichoo S 2024 Drug target interaction prediction using variational quantum classifier *2024 Int. Conf. on Next Generation Computing Applications (NextComp)* (Piscataway, NJ: IEEE) pp 1–7

[21] Paliwal D, Koteswara S N, Gudhanti R, Yadav D and Raj P 2024 Insight into quantum computing and deep learning approach for drug design *Lett. Drug Des. Discov.* **21** 1632–51

[22] Gianelle A, Koppenburg P, Lucchesi D, Nicotra D, Rodrigues E, Sestini L, de Vries J and Zuliani D 2022 Quantum machine learning for *b*-jet charge identification *J. High Energy Phys.* **2022** 14

[23] Wu S L *et al* 2021 Application of quantum machine learning using the quantum kernel algorithm on high energy physics analysis at the LHC *Phys. Rev. Res.* **3** 033221

[24] Incudini M, Bosco D L, Martini F, Grossi M, Serra G and Pierro A D 2024 Automatic and effective discovery of quantum kernels *IEEE Trans. Emerg. Top. Comput. Intell.* 1–10

[25] Meglio A D *et al* 2024 Quantum computing for high-energy physics: state of the art and challenges *PRX Quantum* **5** 037001

[26] Miyabe S *et al* 2023 Quantum multiple kernel learning in financial classification tasks arXiv:2312.00260

**IOP** Publishing

Quantum Machine Learning
Concepts and possibilities
**Andrea Delgado and Kathleen E Hamilton**

# Chapter 5

# Quantum variational optimization

Variational quantum algorithms (VQAs) [1] represent a pivotal shift in how quantum computers can be harnessed for practical problem-solving in the noisy intermediate-scale quantum (NISQ) [2] era. These hybrid algorithms blend the quantum computer's ability to represent and manipulate complex quantum states with the classical computer's strength in performing optimization. The foundational idea is to use a parameterized quantum circuit to prepare a trial quantum state, evaluate an objective function (usually an expectation value), and use a classical optimizer to update the parameters and minimize the cost.

The origins of variational methods trace back to classical quantum chemistry, where the **variational principle** is a fundamental tool for approximating the ground-state energy of a quantum system. This principle states that for any normalized trial wavefunction $|\psi\rangle$, the expectation value of the Hamiltonian $H$ provides an upper bound to the true ground state:

$$E_0 \leqslant \langle\psi|H|\psi\rangle. \tag{5.1}$$

This idea led to the development of methods such as **Hartree–Fock** and **configuration interaction**, where trial wavefunctions are optimized within a parameterized ansatz to approximate molecular eigenstates. These methods became cornerstones of computational chemistry and electronic structure theory throughout the twentieth century.

In the context of quantum computing, the **variational quantum eigensolver (VQE)** was proposed in 2014 by Peruzzo *et al* [3] as a quantum–classical hybrid analog of this traditional variational principle. VQE marked one of the first major algorithms designed specifically for NISQ devices. Instead of representing trial wavefunctions as linear combinations of basis functions (as in Hartree–Fock), VQE uses a quantum circuit with tunable parameters to generate quantum states that approximate the ground state of a given Hamiltonian. This allowed for representing entangled states that are difficult to model classically.

doi:10.1088/978-0-7503-4952-9ch5

The motivation was clear: **quantum chemistry is one of the most demanding applications for classical computers** due to the exponential scaling of the Hilbert space with system size. Even modest molecules can be challenging to simulate using exact diagonalization or classical approximation schemes. Quantum computers, in contrast, can represent these states efficiently, and variational algorithms provide a feasible path to navigate the exponentially large space of quantum states using relatively shallow circuits.

VQE was quickly followed by a wave of algorithmic innovation in variational methods, leading to generalized ansatzes such as the **unitary coupled cluster (UCC)** for more efficient problem solutions and extensions to broader problem classes such as combinatorial optimization with the **quantum approximate optimization algorithm (QAOA)**. Today, variational algorithms are at the center of most practical quantum applications, spanning fields from chemistry and materials science to finance and machine learning.

## 5.1 Model description

Variational quantum algorithms operate by encoding a problem into a cost function that can be evaluated on a quantum device and minimized using classical optimization. The fundamental components of a variational algorithm include a parameterized quantum circuit (ansatz), a quantum measurement procedure to evaluate the cost, and a classical optimization loop that updates the circuit parameters. This hybrid feedback mechanism enables efficient exploration of quantum states even in the presence of hardware limitations.

Let $|\psi(\theta)\rangle = U(\theta)|\psi_0\rangle$ denote the quantum state prepared by a circuit $U(\theta)$, where $\theta$ is a set of real-valued parameters and $|\psi_0\rangle$ is an initial reference state. The goal of a variational algorithm is to find the parameters $\theta$ that minimize a problem-specific cost function, often the expectation value of a Hermitian operator $H$, by evaluating

$$E(\boldsymbol{\theta}) = \langle\psi(\boldsymbol{\theta})|H|\psi(\boldsymbol{\theta})\rangle. \tag{5.2}$$

This cost function is computed by measuring terms in a Hamiltonian decomposition, typically into sums of Pauli operators, and combining them classically.

Two major subclasses of variational algorithms—VQE and QAOA—illustrate the versatility of this framework in both continuous and discrete optimization settings.

### 5.1.1 Variational quantum eigensolver

The VQE was introduced as a practical quantum–classical hybrid algorithm to estimate the ground-state energy of a given Hamiltonian. Its motivation comes directly from the variational principle in quantum mechanics: for any trial state $|\psi\rangle$, the quantity $\langle\psi|H|\psi\rangle$ provides an upper bound on the ground-state energy $E_0$. The closer the trial state is to the true ground state, the lower the energy expectation will be.

In VQE, the Hamiltonian $H$ usually represents the electronic structure of a molecule, expressed in a second-quantized form using fermionic creation and annihilation operators. These operators are mapped to qubit operators via transformations such as the **Jordan–Wigner** [4] or **Bravyi–Kitaev** [5] mappings. The resulting qubit Hamiltonian is a weighted sum of tensor products of Pauli matrices:

$$H = \sum_i c_i P_i, \tag{5.3}$$

where $P_i \in \{I, X, Y, X\}^{\otimes n}$, and $c_i \in \mathbb{R}$ are coefficients determined by the problem Hamiltonian.

To estimate the energy $E(\theta)$, each term $\langle\psi(\theta)|P_i|\psi(\theta)\rangle$ must be measured on the quantum hardware and weighted by $c_i$. The total number of measurements required can be significant, motivating strategies such as **measurement grouping** [6–8], **classical shadows** [9, 10], and **basis rotation grouping** [11] to reduce measurement overhead.

The ansatz $U(\theta)$ plays a central role in VQE's performance. Common choices include:

- **Hardware-efficient ansatzes** [12], which use native gate operations and entangling layers.
- **Problem-inspired ansatzes**, such as the **UCC** [13] ansatz in chemistry.
- **Adaptive ansatzes**, such as **ADAPT-VQE** [14], which grow the circuit based on operator gradients.

The classical optimizer may be derivative-free (e.g. COBYLA [15], Nelder–Mead [16]) or gradient-based (e.g. Adam [17], L-BFGS [18]), depending on whether analytic gradients are available or estimable.

### 5.1.2 Quantum approximate optimization algorithm

The QAOA [19] is a variational algorithm designed to solve combinatorial optimization problems. QAOA operates by encoding the problem into a classical cost function $C(z)$, where $z \in \{0, 1\}^n$ represents a bitstring, and constructing a **cost Hamiltonian** $H_c$ such that its eigenvalues correspond to the cost values for different solutions.

To explore the solution space, QAOA defines a parameterized quantum state by alternating between unitaries generated by the cost and a **mixing Hamiltonian** $H_M$, typically chosen as

$$H_M = \sum_{j=1}^{n} X_j, \tag{5.4}$$

which induces transitions between bitstrings. The QAOA state is then defined as

$$|\psi(\gamma, \beta)\rangle = \prod\nolimits_{j=1}^{p} e^{-i\beta_j H_M} e^{-i\gamma_j H_C}|\psi_0\rangle, \tag{5.5}$$

where $p$ is the depth or number of QAOA layers, and $\gamma, \beta \in \mathbb{R}^p$ are variational parameters.

The initial state $|\psi_0\rangle$ is usually a uniform superposition over all bitstrings, often prepared by applying Hadamard gates to each qubit. The final quantum state

$|\psi(\gamma, \beta)\rangle$ encodes a probability distribution over solutions, and the objective is to maximize the expected value of the cost function by minimizing the expectation

$$\langle\psi(\gamma, \boldsymbol{\beta})|H_C|\psi(\gamma, \boldsymbol{\beta})\rangle. \tag{5.6}$$

QAOA has been applied to various discrete optimization problems, including *MaxCut* [20], *3-SAT* [21], *maximum independent set* [22], and *scheduling* [23]. The theoretical underpinnings relate closely to adiabatic quantum computing, and the performance of QAOA improves with increasing $p$—capable of finding the optimal solution for sufficiently large $p$—though at the cost of circuit complexity and trainability challenges.

## 5.2 Case studies and applications

The flexibility and hybrid structure of variational quantum algorithms have enabled their applications across a diverse array of domains. These applications take advantage of the algorithms' ability to approximate complex quantum states or optimize cost functions that are otherwise classically intractable. Two of the most prominent application domains—quantum chemistry and combinatorial optimization—were in fact the motivating problems for VQE and QAOA, respectively. Since then, the use of variational algorithms has expanded significantly, including into fields such as quantum machine learning, quantum control, and even quantum error mitigation.

### 5.2.1 Quantum chemistry and materials science

Perhaps the most impactful near-term application of variational quantum algorithms lies in simulating molecular systems. The VQE has been used to estimate ground-state energies of small molecules such as $H_2$ [24], LiH [25], $BeH_2$ [26], and $H_2O$ [27], offering benchmarks for early quantum hardware. In these problems, the molecular Hamiltonian is derived from first-principles electronic structure theory and then transformed into a qubit representation via mappings such as the Jordan–Wigner or Bravyi–Kitaev transformations.

These quantum simulations are particularly promising because classical methods such as full configuration interaction (FCI) scale exponentially with system size. VQE offers a polynomial-scaling alternative, assuming that a suitable ansatz can be found and trained. In recent work, adaptive ansatzes and domain-inspired constructions such as UCC methods have improved accuracy while reducing circuit depth. Materials science applications, such as modeling transition metal complexes or strongly correlated systems, are also being explored, although they require higher-fidelity devices and deeper circuits than are currently feasible on most hardware.

### 5.2.2 Combinatorial optimization

The QAOA is specifically designed to tackle classical combinatorial optimization problems, which are ubiquitous in computer science, operations research, logistics, and finance. These problems typically involve optimizing a cost function over discrete variables subject to combinatorial constraints. Examples include:

- **MaxCut:** Finding a partition of a graph that maximizes the number of edges between the two partitions [20, 28].
- **Maximum independent set:** Identifying the largest set of mutually non-adjacent nodes in a graph [22].
- **Vertex cover and graph coloring:** Relevant in resource allocation and scheduling [29, 30].
- **Traveling salesman problem:** A classical route optimization problem [31, 32].

In these applications, the problem is encoded into a diagonal cost Hamiltonian $H_C$, and QAOA generates a parameterized quantum state through alternating applications of $H_C$ and a mixing Hamiltonian $H_M$. Classical post-processing is then used to extract bitstrings corresponding to candidate solutions.

While QAOA has yet to outperform classical algorithms on real-world benchmarks, it has demonstrated potential on small problem instances, in particular in settings where problem-specific structure or initialization strategies can be exploited. Ongoing research into warm-start QAOA, layer-wise training, and optimization landscape analysis aims to improve its scalability and trainability.

### 5.2.3 Quantum machine learning

The variational framework has also been extended to machine learning tasks, where the goal is no longer to find a ground state but rather to train a quantum model to perform classification, regression, or generative modeling. In this context, variational circuits serve as trainable models—analogous to neural networks—whose parameters are updated to minimize a loss function defined over classical or quantum data.

One promising direction is the construction of **quantum classifiers**, where a variational circuit is trained to label classical or quantum inputs. These classifiers can be viewed as quantum analogs of neural networks, with layers of parameterized gates and entangling operations acting as non-linear transformations. The architecture and training of such quantum neural networks will be the subject of the next chapter, chapter 6, where we discuss their structure, expressivity, and role in quantum learning theory.

Another important application is in **quantum generative modeling**, where variational circuits are trained to reproduce the statistical properties of a target distribution. These approaches are particularly well-suited to quantum hardware, as the generative process is inherently probabilistic. We will explore these models in detail in the subsequent chapter, chapter 7, which focuses on generative modeling in quantum machine learning.

Finally, variational circuits also support **quantum kernel methods**, in which a parameterized feature map is used to embed data into a high-dimensional Hilbert space. The inner product between embedded states defines a kernel function, which can be evaluated on a quantum device and used for classical learning tasks such as support vector classification. These methods, including the variational design of quantum feature maps, were discussed in detail in the previous chapter, chapter 4, on information encoding and kernel-based quantum machine learning.

Together, these approaches demonstrate the flexibility of variational quantum algorithms as machine learning tools, capable of encoding both model structure and data representations within a quantum circuit framework.

### 5.2.4 Quantum control and metrology

Variational optimization is also used in quantum control problems, where the objective is to find control parameters that optimize the performance of a quantum system. For example, variational circuits can be used to design pulse sequences that prepare a desired quantum state or maximize measurement sensitivity in quantum metrology set-ups. These applications often use the time-dependent variational principle (TDVP) [33] or other physics-informed cost functions, and they benefit from the same circuit-optimization loop central to VQE and QAOA.

In quantum metrology, variational circuits have been proposed to optimize the sensitivity of quantum sensors [34], particularly in the presence of noise or decoherence. The flexibility of the variational framework makes it possible to tailor sensing strategies to specific hardware constraints or environmental conditions.

### 5.2.5 Quantum simulation and dynamics

Beyond static properties such as ground-state energies, variational algorithms are now being applied to simulate real-time dynamics of quantum systems. The **variational quantum simulation** framework uses time-dependent variational principle to approximate the evolution of quantum states under a given Hamiltonian. This is particularly useful for studying nonequilibrium phenomena in many-body physics, where exact simulation is classically intractable.

Applications in this category include simulation of lattice gauge theories [35], quantum spin models [36], and dissipative systems. These tasks are especially promising for near-term quantum hardware because they may require only modest circuit depth while still offering insights beyond classical reach.

## 5.3 Open challenges and future directions

Despite their promise and relative hardware efficiency, variational quantum algorithms face a number of fundamental and practical **challenges** that limit their scalability and performance. These challenges arise from both quantum hardware constraints and the algorithmic structure of the variational approach itself.

One of the most prominent obstacles is the presence of **barren plateaus** [37] in the optimization landscape. Barren plateaus refer to regions of parameter space where the gradient of the cost function vanishes exponentially with the number of qubits. In such regions, the optimizer receives little to no information about the direction in which to update the parameters, severely impeding the training process. This phenomenon has been shown to occur in both randomly initialized hardware-efficient ansatzes [38] and in certain structured circuits, particularly when global cost functions are used [39]. Strategies to mitigate barren plateaus include the use of local cost functions, shallow-depth circuits, and ansatzes that preserve some form of locality or problem structure. In QAOA, the variational parameters often follow patterns reminiscent of *Trotterized adiabatic evolution*, a structure that has been shown to help mitigate barren plateaus. This connection is well-documented in the literature [40, 41], along with related strategies such as parameter transfer and parameter concentration [42], which enable the reuse of optimized parameters across problem instances—further reducing the need for extensive training and helping to avoid flat regions in the optimization landscape.

Another critical issue is the **selection and expressivity of the ansatz** [43, 44]. The ansatz must be sufficiently expressive to approximate the target state, yet shallow and structured enough to be implementable on NISQ devices. Overly simple ansatzes may not contain the ground state within their reachable subspace, leading to systematic underestimation of the optimal value. Conversely, highly expressive or randomly initialized circuits can suffer from overparameterization and lead to barren plateaus or poor generalization. The design of an effective ansatz is therefore a delicate balance between expressivity, trainability, and hardware efficiency. In chemistry applications, the UCC ansatz provides a physically motivated structure, though its implementation often requires deep circuits. Hardware-efficient ansatzes, while more practical, can lack the inductive bias needed for certain problems.

The **classical optimization step** also presents significant difficulties. The cost function evaluated on the quantum hardware is typically noisy due to finite sampling and hardware imperfections. This noise complicates the performance of gradient-based optimizers and may lead to unstable or inefficient parameter updates [45]. Additionally, the optimization landscape is generally non-convex and full of local minima, saddle points, and flat regions, which can trap classical optimizers and lead to suboptimal solutions. The choice of optimizer and its hyperparameters—such as learning rate or trust region radius—can have a significant impact on the algorithm's performance. Derivative-free methods such as COBYLA or Nelder–Mead are often used in practice, but their performance may degrade with increasing dimensionality.

Another significant challenge is the **measurement overhead** associated with evaluating cost functions. In VQE, for instance, the Hamiltonian is decomposed into a sum of Pauli operators, and the expectation value of each term must be estimated separately through repeated quantum measurements. As the number of terms grows—often polynomially or even exponentially with system size—so too does the number of measurements required for a precise estimate of the energy. This issue is exacerbated by the fact that quantum measurements are probabilistic, requiring many repetitions (shots) to obtain statistically meaningful results. Methods such as measurement grouping, commuting term clustering, and classical shadows have been developed to reduce this burden, but the overhead remains a bottleneck in practice.

Finally, **noise and decoherence** are inherent to NISQ devices and can significantly affect the accuracy of variational algorithms [46, 47]. Quantum gates are not perfectly implemented, and qubits are subject to relaxation and dephasing, in particular during long or deep executions. These effects introduce errors in state preparation and measurement, leading to biased estimates of the cost function and ultimately degraded performance. While variational algorithms are believed to possess some intrinsic noise resilience due to their hybrid structure, in practice the cumulative effect of hardware noise can limit the depth and fidelity of circuits that can be used effectively. Efforts to mitigate these effects include hardware-aware ansatzes, noise-aware optimization, and various error mitigation techniques such as zero-noise extrapolation and probabilistic error cancellation.

Together, these challenges represent a significant research frontier. Understanding and overcoming them is essential to fully realize the potential of variational quantum algorithms in quantum chemistry, combinatorial optimization, and machine learning.

While variational quantum algorithms face a number of obstacles in their practical deployment, they also offer a broad landscape of **opportunities**—both in terms of algorithmic innovation and real-world applications. As researchers develop a deeper understanding of the limitations of NISQ hardware and the expressive capabilities of parameterized quantum circuits, new directions continue to emerge that may unlock scalable quantum advantage.

One particular exciting opportunity lies in the **development of problem-inspired ansatzes**. Instead of relying on generic or hardware-efficient structures, problem-specific circuit designs can leverage the symmetries, structure, or prior knowledge of the physical system being simulated. In quantum chemistry, for example, the UCC ansatz is grounded in well-established methods from classical electronic structure theory and can encode physically meaningful correlations between orbitals. Although the exact implementation of UCC is often too deep for current devices, its truncated and Trotterized variants offer tractable approximations. In combinatorial optimization, analogous opportunities exist in the incorporation of counterdiabatic terms to improve convergence, as well as modifications to the mixing operators that explicitly enforce problem constraints. These enhancements aim to guide the optimization more effectively through the solution space while preserving feasibility [48, 49].

Beyond fixed ansatzes, **adaptive variational algorithms** offer a dynamic approach to circuit construction. Methods such as ADAPT-VQE and operator-pool-based expansions start from a minimal ansatz and iteratively add operators based on gradient information or energy gain heuristics. These adaptive frameworks hold great promise for discovering compact, expressive circuits tailored to the specific instance of a problem, potentially avoiding the overhead associated with overly expressive or randomly initialized circuits. Moreover, by expanding the circuit only when necessary, adaptive approaches help mitigate issues related to barren plateaus and overparameterization.

Variational algorithms are also fertile ground for **advances in quantum-aware optimization strategies**. Unlike traditional machine learning scenarios where gradient computation is efficient and exact, variational quantum algorithms must contend with noisy cost evaluations and stochastic gradients. This has motivated the development of optimization strategies that are robust to quantum noise, such as stochastic parameter shift estimators [50], robust gradient descent [51], and Bayesian optimization [52]. Further opportunities exist in exploring *quantum-aware optimizer scheduling*, where hyperparameters adapt dynamically to noise or signal strength, and in combining classical surrogate models with quantum evaluations to accelerate convergence.

Another important opportunity lies in the **study of expressivity and entanglement in variational circuits**. A growing body of theoretical work seeks to understand the trade-offs between circuit expressivity, trainability, and generalization capacity [53, 54].

Metrics such as expressibility (measuring how well an ansatz explores Hilbert space), entanglement entropy, and Fisher information geometry provide insight into the design of better variational forms [55]. This line of research opens the door to more principled circuit construction, enabling designers to tune expressivity in accordance with the problem scale and noise tolerance.

In parallel, **error mitigation techniques** tailored for variational algorithms are showing strong potential. Because these algorithms require repeated measurements and rely on statistical estimates, they can benefit from emerging mitigation strategies such as zero-noise extrapolation [56], measurement error calibration [57, 58], virtual distillation [7], and Clifford-data regression [59]. Some of these methods are particularly well-suited for variational workflows, as they exploit the structure of repeated circuit evaluations and can be embedded in the training loop.

Finally, perhaps the most exciting opportunity is the **expansion of variational techniques beyond energy minimization**. Although originally designed for finding ground states in quantum chemistry, the variational paradigm has since been applied to a wide range of tasks, including quantum generative modeling, quantum kernel estimation, quantum classifiers, and even quantum control. The flexibility of the variational framework allows for its extension to supervised learning, unsupervised clustering, and reinforcement learning tasks—provided the cost function can be evaluated on a quantum circuit. These developments indicate that variational algorithms may become a foundational tool for near-term quantum machine learning and scientific computing.

As the field matures, the convergence of quantum circuit design, classical optimization, noise-aware methods, and application-specific modeling is likely to push the boundaries of what variational quantum algorithms can achieve. Rather than being a temporary fix for NISQ limitations, these algorithms may evolve into a general-purpose quantum computational paradigm, unlocking new insights across disciplines.

## 5.4 Summary

The versatility of variational quantum algorithms has made them a central pillar of NISQ-era quantum computing. From quantum chemistry to discrete optimization and from machine learning to quantum control, their applications span a wide range of computational domains. While many of these applications remain at the proof-of-concept stage, ongoing improvements in circuit design, optimization techniques, and noise mitigation are rapidly expanding the boundary between theoretical promise and practical utility.

## References


[1] Cerezo M *et al* 2021 Variational quantum algorithms *Nature Rev. Phys.* **3** 625–44

[2] Preskill J 2018 Quantum computing in the NISQ era and beyond *Quantum* **2** 79

[3] Peruzzo A, McClean J, Shadbolt P, Yung M-H, Zhou X-Q, Love P J, Aspuru-Guzik A and O'Brien J L 2014 A variational eigenvalue solver on a photonic quantum processor *Nat. Commun.* **5** 4213

[4] Jordan P and Wigner E 1928 Über das Paulische Äquivalenzverbot *Z. Phys.* **47** 631–51

[5] Bravyi S B and Kitaev A Y 2002 Fermionic quantum computation *Ann. Phys.* **298** 210–26

[6] Yen T-C, Verteletskyi V and Izmaylov A F 2020 Measuring all compatible operators in one series of single-qubit measurements using unitary transformations *J. Chem. Theory Comput.* **16** 2400–9

[7] Huggins W J, McClean J R, Rubin N C, Jiang Z, Wiebe N, Birgitta Whaley K and Babbush R 2021 Efficient and noise resilient measurements for quantum chemistry on near-term quantum computers *npj Quantum Inf.* **7** 23

[8] Gokhale P, Angiuli O, Ding Y, Gui K, Tomesh T, Suchara M, Martonosi M and Chong F T 2019 Minimizing state preparations in variational quantum eigensolver by partitioning into commuting families arXiv: 1907.13623

[9] Huang H-Y, Kueng R and Preskill J 2020 Predicting many properties of a quantum system from very few measurements *Nat. Phys.* **16** 1050–7

[10] Yen T-C, Ganeshram A and Izmaylov A F 2023 Deterministic improvements of quantum measurements with grouping of compatible operators, non-local transformations, and covariance estimates *npj Quantum Inf.* **9** 14

[11] Gonthier J F, Radin M D, Buda C, Doskocil E J, Abuan C M and Romero J 2022 Measurements as a roadblock to near-term practical quantum advantage in chemistry: resource analysis *Phys. Rev. Res.* **4** 033154

[12] Kandala A, Mezzacapo A, Temme K, Takita M, Brink M, Chow J M and Gambetta J M 2017 Hardware-efficient variational quantum eigensolver for small molecules and quantum magnets *Nature* **549** 242–6

[13] Anand A, Schleich P, Alperin-Lea S, Jensen P W K, Sim S, Díaz-Tinoco M, Kottmann J S, Degroote M, Izmaylov A F and Aspuru-Guzik A 2022 A quantum computing view on unitary coupled cluster theory *Chem. Soc. Rev.* **51** 1659–84

[14] Grimsley H R, Economou S E, Barnes E and Mayhall N J 2019 An adaptive variational algorithm for exact molecular simulations on a quantum computer *Nat. Commun.* **10** 3007

[15] Powell M J D 1994 A direct search optimization method that models the objective and constraint functions by linear interpolation *Advances in Optimization and Numerical Analysis* ed S Gomez and J P Hennart (Dordrecht: Springer) pp 51–67

[16] Nelder J A and Mead R 1965 A simplex method for function minimization *Comput. J.* **7** 308–13

[17] Kingma D P and Ba J 2017 Adam: a method for stochastic optimization arXiv: 1412.6980

[18] Byrd R H, Lu P, Nocedal J and Zhu C 1995 A limited memory algorithm for bound constrained optimization *SIAM J. Sci. Comput.* **16** 1190–208

[19] Farhi E, Goldstone J and Gutmann S 2014 A quantum approximate optimization algorithm arXiv: 1411.4028

[20] Wang Z, Hadfield S, Jiang Z and Rieffel E G 2018 Quantum approximate optimization algorithm for MaxCut: a fermionic view *Phys. Rev.* A **97** 022304

[21] Mandl A, Barzen J, Bechtold M, Leymann F and Wild K 2024 Amplitude amplification-inspired QAOA: improving the success probability for solving 3SAT *Quantum Sci. Technol.* **9** 015028

[22] Yang M, Gao F, Wu G, Dai W and Shuang F 2021 A tutorial on quantum approximate optimization algorithm for maximum independent set problem *2021 40th Chinese Control Conf.* (Piscataway, NJ: IEEE) pp 6317–22

[23] Choi J, Oh S and Kim J 2004 Quantum approximation for wireless scheduling arXiv: 2004.11229

[24] Singh D, Mehendale S, Arvind and Dorai K 2024 Ground and excited state energy calculations of the $H_2$ molecule using a variational quantum eigensolver algorithm on an NMR quantum simulator arXiv: 2407.01000

[25] Avramidis B, Paudel H P, Alfonso D, Duan Y and Jordan K D 2024 Ground state property calculations of $LiH_n$ complexes using IBM Qiskit's quantum simulator *AIP Adv.* **14** 035047

[26] Swain K R, Prasannaa V S, Sugisaki K and Das B P 2023 Molecular electric dipole moments: from light to heavy molecules using a relativistic VQE algorithm arXiv: 2211.06907

[27] Kim K, Lim S, Shin K, Lee G, Jung Y, Kyoung W, Kevin Rhee J-K and Min Rhee Y 2024 Variational quantum eigensolver for closed-shell molecules with non-bosonic corrections *Phys. Chem. Chem. Phys.* **26** 8390–6

[28] Zhou Z, Du Y, Tian X and Tao D 2023 QAOA-in-QAOA: solving large-scale MaxCut problems on small quantum machines *Phys. Rev. Appl.* **19** 024027

[29] Tabi Z, El-Safty K H, Kallus Z, Haga P, Kozsik T, Glos A and Zimboras Z 2020 Quantum optimization for the graph coloring problem with space-efficient embedding *2020 IEEE Int. Conf. Quantum Computing Engineering (QCE)* (Piscataway, NJ: IEEE) pp 56–62

[30] Bravyi S, Kliesch A, Koenig R and Tang E 2022 Hybrid quantum-classical algorithms for approximate graph coloring *Quantum* **6** 678

[31] Ramezani M, Salami S, Shokhmkar M, Moradi M and Bahrampour A 2024 Reducing the number of qubits from to $n^2$ to $n\log_2(n)$ solve the traveling salesman problem with quantum computers: a proposal for demonstrating quantum supremacy in the NISQ era arXiv: 2402.18530

[32] Qian W, Basili R A M, Eshaghian-Wilner M M, Khokhar A, Luecke G and Vary J P 2023 Comparative study of variations in quantum approximate optimization algorithms for the traveling salesman problem *Entropy* **25** 1238

[33] Kramer P 2008 A review of the time-dependent variational principle *J. Phys. Conf. Ser.* **99** 012009

[34] Kaubruegger R, Vasilyev D V, Schulte M, Hammerer K and Zoller P 2021 Quantum variational optimization of Ramsey interferometry and atomic clocks *Phys. Rev.* X **11** 041045

[35] Bauer C W *et al* 2023 Quantum simulation for high-energy physics *PRX Quantum* **4** 027001

[36] Chowdhury T A, Yu K, Shamim M A, Kabir M L and Sufian R S 2024 Enhancing quantum utility: simulating large-scale quantum spin chains on superconducting quantum computers *Phys. Rev. Res.* **6** 033107

[37] Larocca M, Thanasilp S, Wang S, Sharma K, Biamonte J, Coles P J, Cincio L, McClean J R, Holmes Z and Cerezo M 2025 Barren plateaus in variational quantum computing *Nat. Rev. Phys.* **7** 174–89

[38] Leone L, Oliviero S F E, Cincio L and Cerezo M 2024 On the practical usefulness of the hardware efficient ansatz *Quantum* **8** 1395

[39] Cerezo M, Sone A, Volkoff T, Cincio L and Coles P J 2021 Cost function dependent barren plateaus in shallow parametrized quantum circuits *Nat. Commun.* **12** 1791

[40] Sack S H and Serbyn M 2021 Quantum annealing initialization of the quantum approximate optimization algorithm *Quantum* **5** 491

[41] Boulebnane S, Sud J, Shaydulin R and Pistoia M 2025 Equivalence of quantum approximate optimization algorithm and linear-time quantum annealing for the Sherrington–Kirkpatrick model arXiv: 2503.09563

[42] Shaydulin R, Lotshaw P C, Larson J, Ostrowski J and Humble T S 2023 Parameter transfer for quantum approximate optimization of weighted maxcut *ACM Trans. Quantum Comput.* **4** 1–15

[43] Du Y, Tu Z, Yuan X and Tao D 2022 Efficient measure for the expressivity of variational quantum algorithms *Phys. Rev. Lett.* **128** 080506

[44] Funcke L, Hartung T, Jansen K, Kühn S and Stornati P 2021 Dimensional expressivity analysis of parametric quantum circuits *Quantum* **5** 422

[45] Larson J, Menickelly M and Shi J 2025 A novel noise-aware classical optimizer for variational quantum algorithms *INFORMS J. Comput.* **37** 63–85

[46] Ito K, Mizukami W and Fujii K 2023 Universal noise-precision relations in variational quantum algorithms *Phys. Rev. Res.* **5** 023025

[47] Wang S, Fontana E, Cerezo M, Sharma K, Sone A, Cincio L and Coles P J 2021 Noise-induced barren plateaus in variational quantum algorithms *Nat. Commun.* **12** 6961

[48] Chandarana P, Hegade N N, Paul K, Albarran-Arriagada F, Solano E, del Campo A and Chen X 2022 Digitized-counterdiabatic quantum approximate optimization algorithm *Phys. Rev. Res.* **4** 013141

[49] Hadfield S, Wang Z, O'Gorman B, Rieffel E G, Venturelli D and Biswas R 2019 From the quantum approximate optimization algorithm to a quantum alternating operator ansatz *Algorithms* **12** 34

[50] Banchi L and Crooks G E 2021 Measuring analytic gradients of general quantum evolution with the stochastic parameter shift rule *Quantum* **5** 386

[51] Stokes J, Izaac J, Killoran N and Carleo G 2020 Quantum natural gradient *Quantum* **4** 269

[52] Dai Z, Ruey Lau G K, Verma A, Shu Y, Hsiang Low B K and Jaillet P 2023 Quantum Bayesian optimization arXiv: 2310.05373

[53] Nakhl A C, Quella T and Usman M 2024 Calibrating the role of entanglement in variational quantum circuits *Phys. Rev.* A **109** 032413

[54] Joch A, Uhrig G S and Fauseweh B 2025 Entanglement-informed construction of variational quantum circuits *Quantum Sci. Technol.* **10** 035032

[55] Haug T and Kim M S 2024 Generalization of quantum machine learning models using quantum Fisher information metric *Phys. Rev. Lett.* **133** 050603

[56] Giurgica-Tiron T, Hindy Y, LaRose R, Mari A and Zeng W J 2020 Digital zero noise extrapolation for quantum error mitigation *2020 IEEE Int. Conf. Quantum Computing Engineering (QCE)* (Piscataway, NJ : IEEE) pp 306–16

[57] Funcke L, Hartung T, Jansen K, Kühn S, Stornati P and Wang X 2022 Measurement error mitigation in quantum computers through classical bit-flip correction *Phys. Rev.* A **105** 062404

[58] Nation P D, Kang H, Sundaresan N and Gambetta J M 2021 Scalable mitigation of measurement errors on quantum computers *PRX Quantum* **2** 040326

[59] Czarnik P, Arrasmith A, Coles P J and Cincio L 2021 Error mitigation with Clifford quantum-circuit data *Quantum* **5** 592

# Chapter 6

# Variational classifiers and neural networks

The fusion of quantum computing and machine learning has led to the emergence of **variational quantum classifiers**. These models leverage quantum circuits to perform classification tasks by optimizing parameters through training. These models sit at the intersection of physics, optimization theory, and computational learning, continuing the historical thread of variational approaches in quantum mechanics that was explored in the previous chapter.

While classical neural networks have undergone decades of development since their conceptual origins in the 1940s with McCulloch and Pitt's formal neuron model [1], the idea of training a quantum circuit for classification is far more recent. However, the inspiration is rooted in a long lineage of ideas: from the variational principles used in chemistry to approximate ground states, to the optimization of circuit parameters for learning tasks in today's noisy intermediate-scale (NISQ) devices.

The **variational quantum circuit (VQC)** as a model of computation emerged in part as a response to the limitations of fully quantum or fully classical models. Early work in this direction, such as the *quantum perceptron* [2] or the *quantum neural network* [3, 4], introduced the idea of learning from data using quantum processes. However, these efforts often lacked a feasible implementation route. It was with the advent of *hybrid quantum–classical algorithms*, notably the variational quantum eigensolver (VQE) and quantum approximate optimization algorithm (QAOA) (discussed in chapter 5), that a practical training loop using quantum circuits became viable.

This chapter picks up from the previous discussion of variational optimization and examines how **hybrid training frameworks** can be adapted for classification tasks. In these settings, a parameterized quantum circuit plays a role analogous to a neural network layer. Input data are encoded into quantum states, transformed through variational gates, and measured to yield predictions. The loss is then

doi:10.1088/978-0-7503-4952-9ch6

computed using classical resources, and parameters are optimized via gradient descent.

The potential advantage of this approach lies in its use of quantum state space as a learning resource. Even with a modest number of qubits, the expressivity of a quantum classifier may exceed that of certain classical models, in particular when using **data re-uploading** [5] or **quantum kernel-inspired architectures**. Yet, practical deployment is far from straightforward. Proof-of-principle experiments on tasks such as image classification and state discrimination can be compared to classical neural network models through analysis of dimensionality and capacity (see chapter 8). However, fully understanding the differences between classical and quantum learning dynamics remains an open question and challenges such as *barren plateaus*, circuit noise, and limited connectivity make the scaling up of quantum neural models a non-trivial question.

In this chapter, we provide a self-contained introduction to **quantum variational classifiers and hybrid neural networks**, describe the training procedure including **quantum-compatible back-propagation and gradient estimation**, and discuss the challenges and opportunities for implementation on current and near-term quantum hardware. This discussion sets the stage for the next chapter, which shifts focus from discriminative models to **quantum generative models** (chapter 7), where circuits are trained to model data distributions.

## 6.1 Model description

In this section, we describe the architecture of variational classifiers by contrasting them with classical neural networks and highlighting how parameterized quantum circuits can be used as building blocks in hybrid learning models.

### 6.1.1 Classical neural networks

Classical neural networks are computational models inspired by the functioning of the human brain. At their core, they consist of simple processing units—artificial neurons—organized in layers (figure 6.1). Each neuron receives inputs, performs a linear combination of them (through weights and biases), and applies a nonlinear activation function to produce its output. This enables the network to model complex, nonlinear relationships between input and output data.

The basic building block of a feed-forward neural network is a **layer** that maps an input vector $\mathbf{x} \in \mathbb{R}^d$ to an output vector $\mathbf{y} \in \mathbb{R}^m$ via

$$\mathbf{y} = \sigma(W\mathbf{x} + \mathbf{b}), \tag{6.1}$$

where:

- $W \in \mathbb{R}^{m\times d}$ is the **weight matrix**,
- $\mathbf{b} \in \mathbb{R}^m$ is the **bias vector**, and
- $\sigma$ is an **element-wise nonlinear activation function**, such as the rectified linear unit (ReLU), sigmoid, or hyperbolic tangent.

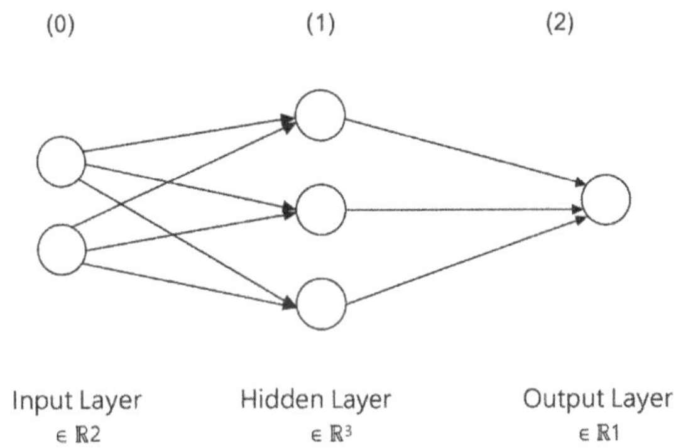


**Figure 6.1.** A simple feed-forward neural network architecture with an input layer in $\mathbb{R}^2$, a hidden layer in $\mathbb{R}^3$ and an output layer in $\mathbb{R}^1$. Each node represents an artificial neuron, and arrows indicate the flow of information through weighted connections. This structure exemplifies how classical neural networks compute nonlinear mappings from input to output via layer-wise transformation.

By stacking multiple such layers, networks are able to learn hierarchical representations of data. For instance, in image classification, the initial layers may learn edges and textures, while deeper layers learn shapes and object categories.

During training, the network parameters $\boldsymbol{\theta} = \{W, \mathbf{b}\}$ are adjusted to minimize a **loss function** $\mathscr{L}(f(\mathbf{x}; \boldsymbol{\theta}), y)$, which quantifies the error between the predicted output and the true label. This is typically done via **gradient descent**, stochastic gradient descent, or one of its many variants. Gradients are computed using **backpropagation**, a recursive application of the chain rule of calculus across the layers of the network.

Neural networks are **universal function approximators** [6, 7], a feed-forward network with a single hidden layer and sufficient units can approximate any continuous function on a compact domain, given the right parameters. This powerful theoretical foundation, combined with efficient training methods and hardware acceleration (e.g. GPUs), has fueled the rise of deep learning across many fields, including computer vision, natural language processing, and scientific data modeling. Universal approximation theorems have also been derived for other neural architectures [8].

### 6.1.2 Parameterized quantum circuits as learning models

As introduced in the previous chapter, a parameterized quantum circuit (PQC) or variational quantum circuit (VQC) consists of an arbitrary set of quantum gates, which are parameterized by a set of controls that determine the outcome of the circuit. A specific type of PQC is so-called quantum neural networks (QNNs), where the suitable set of parameters for a task is learned based on a given set of data, resembling the training of a classical neural network [9–11, 13, 37].

Classical neural networks provide the architectural and conceptual blueprint for variational quantum models. Understanding classical neural networks thus sets the stage for appreciating their quantum counterparts. Quantum circuits, much like neural networks, consist of sequential transformations applied to inputs, followed by a nonlinear output process (in this case, measurement). These gates may be fixed (e.g. Hadamard, CNOT) or parameterized (e.g. rotations $R_X(\theta)$, $R_Y(\theta)$, $R_Z(\theta)$), and their structure defines the **ansatz**, the expressive hypothesis class of the model. The general workflow of a VQC classifier is as follows:

1. **State preparation (data encoding)**: Classical input data $x \in \mathbb{R}^d$ are encoded into a quantum state $|\psi(x)\rangle$ using an **encoding circuit** $U_{\text{enc}}(x)$. This circuit maps the data into the Hilbert space of $n$ qubits.
2. **Parameterized quantum transformation**: A parameterized quantum circuit $U(\theta)$, composed of gates with tunable parameters $\boldsymbol{\theta} = (\theta_1, \theta_2, \ldots, \theta_k)$, is applied to the encoded state. This is the quantum analog of learnable weights in classical neural networks.
3. **Measurement and output**: After applying the variational circuit, a measurement is performed in a chosen basis (usually the computational basis). The classifier output is derived from the **expectation value** of a measurement operator $M$ on the final state,

$$f(x; \boldsymbol{\theta}) = \langle \psi(x; \boldsymbol{\theta}) | M | \psi(x; \boldsymbol{\theta}) \rangle, \tag{6.2}$$

or explicitly in terms of circuit unitaries:

$$f(x; \boldsymbol{\theta}) = \langle 0 | U^{\dagger}_{\text{enc}}(x) \; U^{\dagger}(\boldsymbol{\theta}) \; M \; U(\boldsymbol{\theta}) \; U_{\text{enc}}(x) | 0 \rangle. \tag{6.3}$$

The result of a measurement can be interpreted in different ways, depending on the classification task:

- **Binary classification:** A single-qubit observable (e.g. $Z$) is measured, and the sign of the expectation value is mapped to class labels.
- **Multiclass classification:** Multiple qubits are measured to produce logits for each class, or a one-versus-all strategy is used.

The training loop resembles classical training, but with quantum evaluations at the forward step and specialized techniques (e.g. parameter-shift rule) for computing gradients. The final state before measurement is

$$|\psi(x; \boldsymbol{\theta})\rangle = U(\boldsymbol{\theta}) \; U_{\text{enc}}(x) \; |0\rangle^{\otimes n}. \tag{6.4}$$

This hybrid structure *classical input → quantum embedding → parameterized quantum evolution → classical output* forms the backbone of quantum classification models.

Like neural networks, VQCs can be stacked, extended with data re-uploading layers, and used in conjunction with classical layers. However, their training dynamics and representational capacity differ significantly, motivating new theoretical and empirical analysis.

## 6.2 Backpropagation and gradient estimation in quantum models

Training variational quantum classifiers requires **differentiating expectation values of observables with respect to circuit parameters**. Unlike classical neural networks, where backpropagation flows through a graph of differential primitives, quantum hardware only returns **samples from measurement outcomes**, so gradients must be estimated through **hardware-compatible rules**. In hybrid models, these quantum gradients are then stitched into a larger *end-to-end* auto-differentiation pipeline. Below we survey the main techniques, their assumptions, costs, and when to use them.

### 6.2.1 Automatic differentiation on hybrid computational graphs

In modern machine learning, automatic differentiation (autodiff) [14] is the backbone of efficient gradient-based training. Whether in PyTorch [15], TensorFlow [16], or JAX [17], autodiff enables seamless computation of derivatives by constructing and traversing a **computational graph**. In the context of quantum machine learning, the introduction of hybrid models—where quantum circuits are embedded within classical learning architectures—raises a crucial question: how do we propagate gradients through a model that contains a quantum circuit?

The answer lies in **hybrid computational graphs**, which extend the autodiff paradigm to include *quantum nodes*. These nodes encapsulate quantum operations (state preparation, variational evolution, and measurement) and expose *custom gradient rules* so that classical optimizers can update parameters across the quantum–classical boundary.

**Hybrid graph structure**

A typical hybrid mode consists of:

- **Classical nodes:** feature pre-processing, dense layers, and loss functions.
- **Quantum nodes:** parameterized quantum circuits acting on input-encoded states, returning expectation values or sample statistics.

These nodes are linked in a computation graph, where data flow **forward** during evaluation, and gradients flow **backward** during training. The overall architecture resembles a classical deep learning pipeline, but with variational quantum circuits taking the role of trainable nonlinear transformations.

Formally, let $\boldsymbol{x} \in \mathbb{R}^d$ be a $d$-dimensional classical feature vector, and let $|\psi(\boldsymbol{x};\, \boldsymbol{\theta})\rangle$ be the state prepared by the quantum circuit parameterized by the vector of angles $\boldsymbol{\theta}$. The output of the quantum node is $f(\boldsymbol{x};\, \boldsymbol{\theta})$—for example, if the output is an expectation value of an operator $M$:

$$f(x;\, \boldsymbol{\theta}) = \langle \psi(x;\, \boldsymbol{\theta}) | M | \psi(x;\, \boldsymbol{\theta}) \rangle. \tag{6.5}$$

The output is a scalar value bounded between $[-1,\, 1]$ from which a label $\hat{y}(x;\, \boldsymbol{\theta})$ is extracted. Examples of label extraction include using a classical bias parameter to assign binary labels, or re-scaling values from $[-1,\, 1]$ to $[0,\, 1]$. A classical loss function $\mathscr{L}(y,\, \hat{y})$ is then evaluated using ground truth label $y$. The gradient with respect to the quantum parameters is

$$\nabla_{\theta}\ \mathscr{L}(f(x;\, \boldsymbol{\theta}),\, y) = \frac{\partial \mathscr{L}}{\partial f} \cdot \frac{\partial f}{\partial \boldsymbol{\theta}}. \tag{6.6}$$

Here, $\frac{\partial \mathscr{L}}{\partial f}$ is computed classically, while $\frac{\partial f}{\partial \theta}$ must be estimated using quantum-compatible methods, such as the parameter-shift rule.

**Frameworks enabling hybrid auto-differentiation**

Several quantum machine learning frameworks provide infrastructure for hybrid auto-differentiation:

- **PennyLane** [18] enables full differentiable quantum–classical graphs, supporting PyTorch, TensorFlow, and JAX backends. It exposes *quantum nodes* with built-in gradient rules and supports end-to-end optimization with standard optimizers.
- **Qiskit Machine Learning** [19] integrates with PyTorch via custom TorchConnector modules and supports parameter-shift gradients or finite-difference estimators on IBM backends or simulators.
- **TensorFlow Quantum** [20] supports differentiable quantum layers integrated with classical TensorFlow/Keras pipelines. It uses adjoint differentiation in simulation and sampling-based gradient rules on hardware.

These frameworks automatically keep track of which parts of the graph are quantum versus classical, and invoke the correct gradient rules accordingly.

**Example: Hybrid training loop**

Consider a hybrid model where a quantum circuit produces a real-valued output $f(x; \boldsymbol{\theta})$ used in a logistic regression task. The loss might be

$$\mathscr{L}(f, y) = -y \log(f) - (1 - y)\log(1 - f). \tag{6.7}$$

During training:

- The **forward pass** evaluates the quantum circuit on input $x$, measures the observable, and computes the scalar output $f$.
- The loss $\mathscr{L}$ is computed and stored in the graph.
- The **backward pass** starts from $\nabla_f \mathscr{L}$, uses quantum gradient rules to evaluate $\nabla_\theta(f)$, and applies the chain rule to compute $\nabla_\theta \mathscr{L}$.
- A classical optimizer (e.g. Adam) updates the parameters $\boldsymbol{\theta}$.

The entire process is transparent to the user, who simply defines the model and loss in a declarative way, much like building a classical neural network.

**Differentiability in practice**

Differentiability across the quantum–classical boundary is possible only because quantum circuits can be expressed as differentiable expectation-value functions, given suitable gates and observables. However, differentiability **does not mean analyzability** due to sampling noise, barren plateaus, and hardware limitations; care must be taken to ensure gradients are informative and reliable.

Nevertheless, hybrid autodiff has made variational quantum models practical. It enables the integration of quantum modules into large-scale machine learning systems, facilitating experimentation with architectures where quantum computation plays a meaningful, trainable role.

### 6.2.2 Parameter-shift rule

One of the main break-throughs enabling the training of variational quantum circuits on quantum hardware is the **parameter-shift rule (PSR)** [21]. This rule allows us to compute the gradient of an expectation value with respect to a circuit parameter, using only evaluations of the quantum circuit at shifted values of the parameter, meaning no access to the internal quantum state or wavefunction is required. This makes the PSR a **hardware-friendly and analytically exact** method for computing gradients on NISQ devices.

**Mathematical derivation**

Suppose we have a parameterized quantum circuit $U(\theta)$ composed of gates where a specific gate $R(\theta)$ has the form

$$R(\theta) = e^{-i\theta G}, \tag{6.8}$$

where $G$ is a Hermitian generator with two distinct eigenvalues $\pm r$ (common case: $r = \frac{1}{2}$, as in Pauli rotations).

Let the circuit prepare the quantum state,

$$|\psi(\theta)\rangle = U(\theta)|\psi_0\rangle, \tag{6.9}$$

and let the output be the expectation value of some observable $M$:

$$f(\theta) = \langle\psi(\theta)|M|\psi(\theta)\rangle. \tag{6.10}$$

Then the **parameter-shift rule** states

$$\frac{\partial f}{\partial \theta} = \frac{1}{2r}\left[f\left(\theta + \frac{\pi}{4r}\right) - f\left(\theta - \frac{\pi}{4r}\right)\right], \tag{6.11}$$

where $s = \frac{\pi}{4r}$.

For the common case where $G = \frac{1}{2}Z$ (i.e. single-qubit rotation gates such as $R_X$, $R_Y$, $R_Z$), this simplifies to

$$\frac{\partial f}{\partial \theta} = \frac{1}{2}\left[f\left(\theta + \frac{\pi}{2}\right) - f\left(\theta - \frac{\pi}{2}\right)\right]. \tag{6.12}$$

This result is exact and does not involve finite-difference approximations or knowledge of the wavefunction, only evaluations of the same circuit with different parameter values.

**Intuition**

In variational circuits, parameterized gates are often single qubit rotations about the ($X$, $Y$, and $Z$) axes:

- $R_X(\theta) = e^{-i\theta X/2}$,
- $R_Y(\theta) = e^{-i\theta Y/2}$,
- $R_Z(\theta) = e^{-i\theta Z/2}$.

The parameter-shift rule tells us that the effect of changing $\theta$ on the measurement outcome can be captured by evaluating the circuit at two symmetrically shifted points. This is particularly advantageous because it works even when the circuit is run on a noisy quantum device.

**Practical implementation**

To compute the gradient of a loss function with respect to a parameter $\theta_i$, one must:

1. Shift $\theta_i$ forward by $\frac{\pi}{2}$, evaluate the circuit, and record the measurement outcome.
2. Shift $\theta_i$ backward by $\frac{\pi}{2}$, evaluate the circuit again.
3. Take the difference, divide by 2, and chain with the gradient of the loss function.

This results in **two forward passes per trainable parameter**.

**Example: Parameter-shift rule**

Suppose we are measuring a qubit in the $Z$ basis after applying a rotation $R_Y(\theta)$. Then the expectation value is

$$f(\theta) = \langle 0|R_Y^{\dagger} Z R_Y|0\rangle = \cos(\theta), \tag{6.13}$$

so

$$\frac{\partial f}{\partial \theta} = -\sin(\theta) = \frac{1}{2}\left[\cos\left(\theta + \frac{\pi}{2}\right) - \cos\left(\theta - \frac{\pi}{2}\right)\right]. \tag{6.14}$$

This confirms the PSR for this case.

**Limitations**

The parameter-shift rule applies only to gates with generators that have **two unique eigenvalues**, typically $\pm r$. When the generator has more then two eigenvalues (e.g. controlled unitaries or non-Pauli rotations), the rule must be generalized.

Furthermore, while PSR is exact, it can be **computationally expensive** when the number of parameters is large. For a circuit with $p$ parameters, computing the full gradient requires $2p$ circuit evaluations per data point, each run subject to shot noise if evaluated on real hardware.

### 6.2.3 Generalized parameter-shift rules

While the standard PSR applied cleanly to gates with generators having only two distinct eigenvalues, many quantum operations of interest have **generators with more than two distinct eigenvalues**. In such cases, the original PSR must be extended.

The **generalized parameter-shift rule** [22] offers a framework for computing derivatives of expectation values for arbitrary gates of the form

$$U(\theta) = e^{-i\theta G}, \tag{6.15}$$

where $G$ is a Hermitian operator with an arbitrary spectrum spec($G$) = $\{g_1, g_2, \dots, g_k\}$.

**General form of the gradient**

Given a quantum circuit that prepares the state $|(\psi)\rangle = U(\theta)|\psi_0\rangle$, and an observable $M$, the expectation value is

$$f(\theta) = \langle\psi(\theta)|M|\psi(\theta)\rangle. \tag{6.16}$$

Then, under suitable regularity conditions (e.g. that $G$ has a discrete spectrum and the circuit is differentiable), the derivative $f(\theta)$ can be expressed as a finite weighted sum of circuit evaluations at **multiple shifted parameter values,**

$$\frac{\partial f}{\partial \theta} = \sum\nolimits_{s\in\mathscr{F}} c_s \cdot f(\theta + s), \tag{6.17}$$

where:

- $\mathscr{F} \subset \mathbb{R}$ is a set of shift values and
- $c_s \in \mathbb{R}$ are coefficients depending on the spectrum of $G$.

The set $\{c_s, s\}$ is obtained via **Fourier analysis** of the generator's spectral decomposition. Importantly, this approach still only requires forward evaluations of the quantum circuit, making it compatible with hardware.

**Key observations**

- The number of shifts required depends on the **number of distinct eigenvalues** of $G$. For Pauli generators (two eigenvalues), two shifts suffice; for more complex gates, the number of required shifts increase accordingly.
- For **multi-qubit gates** such as controlled-phase or controlled-rotation gates, the generator circuit often has four or more eigenvalues. In this case, one may need four, six, or even more circuit evaluations to estimate the gradient of a single parameter.
- **Diagonal generators** (e.g. projectors or Hamiltonian terms in block-diagonal form) can often be decomposed analytically, facilitating an efficient generalized PSR formulation.

**Example: Three eigenvalue generator**

Suppose $G$ has eigenvalues $\{-1, 0, +1\}$. Then the generalized PSR yields:

$$\frac{\partial f}{\partial \theta} = \sum\nolimits_k c_k f(\theta + s_{k,}) \tag{6.18}$$

with shift values $s_k \in \{\pm\pi/2, \pm\pi/4\}$, and coefficients $c_k$ computed to match the derivative of $f$. The specific shifts and weights depend on the gate decomposition and the operator $M$, but remain computable for a wide class of circuits.

**Trade-offs**
While generalized PSR preserves many of the advantages of the original rule—such as no acillae, hardware-compatibility, and bias-free estimation—it introduces a few complications:

- **Increased cost:** The number of evaluations per parameter scales with the number of unique eigenvalues of the generator. In the worst case, this can become impractical.
- **Complex precomputation:** The coefficients $c_s$ and shifts $s$ are model-dependent and must be derived for each generator.
- **Less common in practice:** Most near-term applications are still dominated by circuits from Pauli rotations and controlled-NOT gates, where the standard PSR suffices.

However, as quantum circuits grow more expressive and incorporate more diverse gate sets (e.g. in analog quantum computing or Hamiltonian simulation), generalized PSR will likely become increasingly relevant.

**Alternative: gate decomposition**
In practice, rather than applying the generalized rule directly, it is often preferable to decompose complex gates into standard gates with known parameter-shift rules. For example:

- Controlled-rotation gates can be decomposed into basic Pauli rotations plus CNOTs.
- Multi-qubit unitaries can often be decomposed into a universal gate set (e.g. Clifford + T), enabling use of the basic PSR.

This strategy reduces the burden of developing shift rules for each new gate, at the cost of increasing circuit depth.

### 6.2.4 Finite differences and simultaneous perturbation stochastic approximation

While the PSR provides exact, hardware-compatible gradients under specific conditions, there are scenarios where it may not be applicable or practical:

- The generator of a gate has a complex spectrum or is unknown.
- The gate is non-analytic or device-specific.
- The cost of multiple circuit evaluations per parameter is too high.

In such cases, **black-box gradient estimators** such as **finite-differences** and **simultaneous perturbation stochastic approximation (SPSA)** [23] offer useful alternatives. These methods do not rely on circuit structure or spectral properties and can be used to train any differentiable or non-differentiable function approximator, including noisy or proprietary quantum hardware.

#### *6.2.4.1 Finite-difference estimators*
Finite-difference methods estimate the gradient of a function by evaluating the function at nearby points and computing a difference quotient. For a scalar function $f(\theta)$, the **central finite-difference** estimate is

$$\frac{\partial f}{\partial \theta} \approx \frac{f(\theta + \varepsilon) - f(\theta - \varepsilon)}{2\varepsilon}. \tag{6.19}$$

This method is conceptually simple and model-agnostic. It requires only two forward evaluations per parameter, and no assumptions about the functional for of $f$. In vector-valued models, gradients are estimated **parameter-by-parameter**, making the cost scale linearly with the number of parameters.

**Pros:**

- Very easy to implement.
- Works for any differentiable or even non-differentiable model.
- Requires only access to forward evaluations (black-box compatible).

**Cons:**

- Introduces bias controlled by the step size $\varepsilon$; small $\varepsilon$ reduces bias but increases variance due to noise.
- Computational cost scales poorly with the number of parameters.
- Sensitive to hardware noise and sampling fluctuations.

*6.2.4.2 Simultaneous perturbation stochastic approximation*

SPSA is a stochastic gradient approximation algorithm that estimates the full gradient using **only two function evaluations**, regardless of the number of parameters. This is achieved by perturbing all parameters simultaneously along a random direction and estimating the gradient via a finite-difference quotient.

Let $\boldsymbol{\theta} \in \mathbb{R}^p$ be the parameter vector. At each iteration $k$, we sample a random direction $\Delta_k \in \{-1, +1\}^p$, then evaluate the function at

$$\boldsymbol{\theta}^{\pm} = \boldsymbol{\theta} \pm c_k \boldsymbol{\Delta}_k, \tag{6.20}$$

and estimate the gradient as

$$\hat{g}_i^{(k)} = \frac{\mathscr{L}(\boldsymbol{\theta}^+) - \mathscr{L}(\boldsymbol{\theta}^-)}{2c_k(\Delta_k)_i}. \tag{6.21}$$

**Key features**

- Requires two circuit evaluations per training step, regardless of parameter count.
- Particularly useful in high-dimensional optimization tasks or where evaluating gradients directly is expensive.
- Can be used with noisy or partially inaccessible quantum hardware, including analog devices.

**Pros:**

- Dimension-independent cost (two evaluations per update).
- Naturally noise-tolerant and robust to gradient sparsity.
- Suitable for noisy hardware, black-box models, and non-differentiable settings.

**Cons:**

- Gradient estimates have high variance.
- Convergence is slower compared to exact gradients.
- Often requires tuning of learning rate $a_k$ and perturbation size $c_k$.

*6.2.4.3 Practical considerations*

- Use finite differences when PSR is unavailable but you are working with simulators or moderately noisy devices where variance is controlled.
- Use SPSA when you are dealing with hardware noise, high-dimensional models, or you need a cheap estimate of the gradient direction.

Both methods are compatible with gradient-free optimization as well, meaning you can also use them to drive derivative-free optimizers (e.g. COBYLA, Nelder–Mead, CMA-ES).

*6.2.4.4 Comparison with parameter-shift*

Table 6.1 shows a comparison between the gradient estimation methods discussed in this section in terms of number of evaluations, and some caveats to consider when selecting the appropriate method for a given application.

**Table 6.1.** Comparison of gradient estimation methods used in training variational quantum circuits.

| Method | Exact? | # Eval/ param. | Total eval. | Dim.-agnostic | Notes |
|---|---|---|---|---|---|
| Parameter-shift | Yes | 2 | $2p$ | No | Preferred when PSR applies; low variance |
| Finite-difference | No | 2 | $2p$ | No | Simple, biased; sensitive to $\varepsilon$ and noise |
| SPSA | No | — | 2 | Yes | High variance; efficient in high dimensions and on noisy hardware |

### 6.2.5 Adjoint differentiation (simulators only)

Adjoint differentiation [24], also referred to as reverse-mode differentiation for quantum circuits, is a gradient computation technique that is exact, but it is only available in state-vector simulators. Unlike the PSR, which estimates gradients by evaluating the circuit multiple times with shifted parameters, the adjoint method propagates derivative analytically and backward through the circuit, akin to classical backpropagation.

This method is **not compatible with real quantum hardware** because it requires full access to the quantum state, including the ability to apply adjoint operations and compute overlaps between intermediate states, which are quantities that are inaccessible on physical devices due to the no-cloning theorem and measurement collapse.

*6.2.5.1 Conceptual overview*

Let a quantum circuit $U(\theta)$ prepare a state,

$$|\psi(\theta)\rangle = U(\theta)|0\rangle^{\otimes n}, \tag{6.22}$$

and let the output of the model be the expectation value of a Hermitian observable $M$,

$$f(\boldsymbol{\theta}) = \langle\psi(\boldsymbol{\theta})|M|\psi(\boldsymbol{\theta})\rangle. \tag{6.23}$$

The goal is to compute the gradient $\nabla_\theta f$. Using adjoint differentiation, this is done by propagating gradients backwards through each gate in the circuit similar to classical neural backpropagation.

This approach leverages the **chain rule for unitary gates**, and computes each partial derivative with respect to a gate parameter $\theta_i$ using

$$\frac{\partial f}{\partial \theta_i} = 2\ \mathrm{Re}\left\langle \frac{\partial\psi(\boldsymbol{\theta})}{\partial \theta_i} |M|\psi(\boldsymbol{\theta})\right\rangle. \tag{6.24}$$

Critically, the derivative $\partial|\psi\rangle/\partial\theta_i$ is not estimated numerically but computed exactly by **catching intermediate states and applying the adjoint (reverse) of each gate**, just like reverse-mode autodiff in classical machine learning.

*6.2.5.2 Efficiency and scaling*

One of the strengths of adjoint differentiation is its computational efficiency. For quantum circuits with $G$ gates and $p$ parameters, the full gradient can be computed in $\mathcal{O}(G)$ time and memory, rather than $\mathcal{O}(p)$ evaluations as in parameter-shift. This makes it particularly appealing for:

- Large variational circuits with many parameters,
- Batch-mode optimization during model prototyping, and
- Architecture searches and hyperparameter sweeps where fast gradients are needed.

*6.2.5.3 Limitations*

Despite its utility, adjoint differentiation is not usable on hardware. It requires:

- **Access to the full quantum state**, which is exponentially large in the number of qubits.
- **Reversible simulation of gates**, including full complex amplitudes.
- **Post-selection or overlap measurements**, which cannot be implemented without cloning or tomography.

Additionally, memory usage scales linearly with the number of gates (due to backpropagation through cached states), and cannot scale well beyond 30–40 qubits in practice.

### 6.2.6 Quantum natural gradient and the quantum geometric tensor

While traditional gradient descent is widely used in both classical and quantum machine learning, it is not always the most efficient path to optimization. This is especially true in quantum models, where the parameter space is often curved due to the geometry of quantum states. In such cases, standard gradients can be misaligned with the natural directions of change in the state space, leading to poor convergence.

The **quantum natural gradient (QNG)** [25] corrects this by accounting for the **Riemannian geometry** of the quantum state manifold. It generalizes the classical notion of the natural gradient—used in statistical learning to improve convergence by adjusting for curvature—by incorporating quantum-specific metrics derived from information geometry.

*6.2.6.1 From Euclidean to Riemannian geometry*

In ordinary gradient descent, the update rule is

$$\boldsymbol{\theta} \leftarrow \boldsymbol{\theta} - \eta \, \nabla_{\boldsymbol{\theta}} \, \mathscr{L}, \tag{6.25}$$

where $\eta$ is the learning rate and the gradient is taken with respect to the Euclidean geometry of parameter space.

However, this approach does not account for how a change in parameters affects the quantum state, i.e. it ignores the fact that two different parameter updates might lead to very different magnitudes of change in the underlying state $|\psi(\theta)\rangle$.

The quantum natural gradient modifies the update rule to follow the geodesics of the quantum state by rescaling the gradient using the inverse of the quantum Fisher information metric [26, 27], which is equivalent to the real part of the quantum geometric tensor (QGT):

$$\dot{\boldsymbol{\theta}} = -\eta \, F^{-1}(\boldsymbol{\theta}) \;\; \nabla_{\boldsymbol{\theta}} \, \mathscr{L}. \tag{6.26}$$

*6.2.6.2 The quantum geometric tensor*

The QGT $F_{ij}(\theta)$ captures how sensitive the quantum state is to changes in the parameters. It is defined as

$$F_{ij}(\boldsymbol{\theta}) = \langle \partial_i \psi | \partial_j \psi \rangle - \langle \partial_i \psi | \psi \rangle \langle \psi | \partial_j \psi \rangle. \tag{6.27}$$

The **real part** of this quantity defines the **quantum Fisher information matrix**, which is used in the QNG:

$$\mathrm{Re}\left[ F_{ij} \right] = \mathrm{Cov}\left( \frac{\partial}{\partial \theta_i}, \frac{\partial}{\partial \theta_j} \right). \tag{6.28}$$

The QGT behaves like a **local metric tensor**, measuring how far apart infinitesimal parameter changes move to the quantum state in the **Fubini–Study metric** (which governs distance in projective Hilbert space).

*6.2.6.3 Why it matters*

- In regions of parameter space where the quantum state is **insensitive** to parameter changes (i.e. the landscape is flat), the natural gradient **boosts** updates.
- In regions where the state is **highly sensitive**, the update is scaled down to prevent overshooting.

This **geometry-aware learning** leads to:

- Faster convergence,
- More stable optimization, in particular in the presence of barren plateaus, and
- Better conditioning in circuits with high parameter redundancy.

*6.2.6.4 Computation and implementation*

In practice, computing $F^{-1}(\theta)$ can be expensive:

- The QGT is a $p \times p$ matrix, where $p$ is the number of parameters.
- Each entry requires evaluating overlaps between quantum state derivatives.

To address this, common strategies include:

- Using diagonal or block-diagonal approximations,
- Replacing $F^{-1}$ with a low-rank update (e.g. from recent parameter history), and
- Estimating $F$ on a subset of training data or using empirical Fisher approximations.

Some quantum software libraries such as PENNYLANE and QISKIT provide utilities for computing or approximating the QGT, often via parameter-shift rules for QGT entries.

### 6.2.7 Comparison to the classical natural gradient

Table 6.2 highlights the differences between the classical and quantum natural gradient methods for the optimization of quantum circuits in terms of their features.

**Table 6.2.** Comparison between classical and quantum natural gradient methods for optimization.

| Feature | Classical natural gradient | Quantum natural gradient |
|---|---|---|
| Metric | Fisher information (probability distribution) | Quantum Fisher information (QGT) |
| Geometry | Euclidean in parameter space | Fubini–Study in Hilbert space |
| Use case | Deep learning, variational inference | Variational quantum circuits, QML |
| Computation cost | Moderate | High (matrix inversion + overlap estimation) |
| Scalability | Works well for large-scale ML models | Challenging for large quantum circuits |
| Implementation | Supported in many classical ML libraries | Available in select QML toolkits (e.g. PennyLane, Qiskit) |

### 6.2.8 Gradients for mixed states, noisy channels, and open-system dynamics

So far, we have discussed gradient estimation under the assumption that quantum computations involve **pure states** evolving unitarily. However, real-world quantum systems are **open**, interacting with their environments through noise and decoherence. Additionally, some applications such as variational algorithms for open-system simulation or error-mitigated quantum learning explicitly involve **quantum channels** and **mixed states**.

In this section, we explore how to estimate gradients when the output state is mixed or evolves under a non-unitary map, and how standard techniques such as the parameter-shift rule can be extended or replaced in these cases.

*6.2.8.1 Mixed states and Kraus operators*

A mixed quantum state is described by a density matrix $\rho(\theta)$, and its evolution can be modeled using **quantum channels** $\varepsilon$, which are completely positive and trace-preserving (CPTP) maps.

The general form of a channel acting on a state is

$$\rho(\boldsymbol{\theta}) = \sum_k K_k(\boldsymbol{\theta})\rho_0 K_k^\dagger(\boldsymbol{\theta}), \tag{6.29}$$

where $\{K_k(\theta)\}$ are the **Kraus operators** of the channel, and may depend on variational parameters.

The observable expectation value becomes

$$f(\boldsymbol{\theta}) = \mathrm{Tr}[M\rho(\boldsymbol{\theta})] = \sum_k \mathrm{Tr}[MK_k(\boldsymbol{\theta})\rho_0 K_k^\dagger(\boldsymbol{\theta})]. \tag{6.30}$$

The gradient of this function can be computed by differentiating the Kraus operators:

$$\frac{\partial f}{\partial \theta} = \sum_k \mathrm{Tr}\left[\frac{\partial K_k}{\partial \theta}\rho_0 K_k^\dagger M + K_k\rho_0\frac{\partial K_k^\dagger}{\partial \theta}M\right]. \tag{6.31}$$

This expression is exact but generally requires **access to the functional form** of the Kraus operators.

*6.2.8.2 Noisy quantum circuits*

When circuits are executed on hardware, noise acts between gates or at measurement. This leads to effective states of the form

$$\rho_{\mathrm{out}}(\theta) = \mathcal{N}_L{}^\circ\mathcal{U}_L(\theta_L)^\circ. \ldots\mathcal{N}_1{}^\circ\mathcal{U}_1(\theta_1)(\rho_0), \tag{6.32}$$

where:

- $\mathcal{U}_i$ are unitary gates (parameterized) and
- $\mathcal{N}_i$ are noise channels (e.g. depolarizing, amplitude damping).

In this context, gradient estimation methods that assume pure states become inaccurate. Shot noise, stochastic gate noise and hardware drift all contribute to gradient variance and bias. To combat this:

- **Finite-difference and SPSA** remain valid, as they make no assumption about the internal circuit.

- **Noisy parameter-shift rules** may still apply if the noise is gate-independent or can be modeled analytically.
- **Noise-aware adjoint methods** are being developed, but require simulation of the noise models.

*6.2.8.3 Parameter-shift for channels*

Recent work [28] has shown that **parameter-shift rules can be extended to channels**, provided that the channel can be decomposed as

$$\varepsilon(\theta)[\rho] = A(\theta)\rho A^{\dagger}(\theta) + B(\theta)\rho B^{\dagger}(\theta) \tag{6.33}$$

and the operators $A(\theta)$, $B(\theta)$ admit shift rules similar to those of unitary gates. Under certain conditions (e.g. commuting with a Pauli basis), parameter-shift rules still hold. However, in general, one must:

- Use finite differences for gradients,
- Estimate derivatives using Monte Carlo techniques, and
- Purify the state and simulate it on an extended Hilbert space.

*6.2.8.4 Purification-based gradients*

Any mixed state $\rho(\theta)$ can be represented as a pure state on a larger Hilbert space via purification:

$$\rho(\boldsymbol{\theta}) = \mathrm{Tr}_{\mathrm{anc}}[|\Psi(\boldsymbol{\theta})\rangle\langle\Psi(\boldsymbol{\theta})|]. \tag{6.34}$$

By constructing a unitary circuit that prepares this purified state on $\mathscr{H}_{\mathrm{system}} \otimes \mathscr{H}_{\mathrm{ancilla}}$, we can apply standard parameter-shift rules to estimate gradients. However, this comes at the cost of doubling the qubit requirement, increasing circuit depth, and introducing additional post-processing overhead (e.g. tracing out the ancilla subsystem).

### 6.2.9 Gradient variance, shot noise, and batching

Unlike classical computation, where evaluating a function or its gradient returns a deterministic value (up to floating-point noise), quantum computations are inherently probabilistic. When measuring quantum states, we obtain discrete outcomes sampled from a probability distribution defined by the quantum state and measurement observable. Consequently, expectation values and their derivatives must be estimated via repeated measurements, also known as **shots**.

This introduces **variance** in both forward and backward passes of a quantum model. Understanding and managing this variance is crucial for the stability and efficiency of variational training.

*6.2.9.1 Shot noise and expectation estimation*

Suppose a variational quantum circuit returns an expectation value

$$f(\boldsymbol{\theta}) = \mathrm{Tr}[M\rho(\boldsymbol{\theta})], \tag{6.35}$$

but we can only access it through **shot-based sampling** on a quantum device. If $M$ has eigenvalues in $[-1, 1]$, then the **standard deviation** of the sample mean over $N$ shots is

$$\mathrm{Std}[\hat{f}\,] = \frac{\sqrt{\mathrm{Var}(M)}}{\sqrt{N}} \leqslant \frac{1}{\sqrt{N}}. \tag{6.36}$$

Thus, to reduce the error in the estimate by a factor of 10, we must increase the number of shots by a factor of 100. This statistical noise is referred to as **shot noise**, and it affects:

- **Forward evaluations** of the model output,
- **Gradient estimates**, in particular when using the parameter-shift rule, and
- **Loss values**, which are often nonlinear functions of noisy outputs.

*6.2.9.2 Gradient variance and signal-to-noise ratio*

Let $\hat{g}_j$ be the estimated gradient of parameter $\theta_j$. The variance of this estimate scales with:

- The number of shots,
- The functional sensitivity of the circuit, and
- The parameterization of the gates.

When gradients are small (e.g. near flat regions or in barren plateaus), **shot noise can dominate**, making the signal indistinguishable from noise. This leads to:

- Unstable optimization steps,
- Poor convergence or stagnation, and
- Bias in gradient direction due to measurement error.

Quantifying the **signal-to-noise ratio (SNR)** of gradients is essential in practice. The **Fisher information** or **gradient norm** can serve as proxies for when to adapt shot counts or learning rates.

*6.2.9.3 Adaptive shot allocation*

To mitigate the impact of shot noise:

- Use **adaptive shot strategies** [29, 30]: allocate more shots to parameters or data points with high gradient variance.
- Estimate the **variance per parameter** during training and dynamically adjust measurement budgets.
- Use **variance-reduced estimators**, e.g. averaging over symmetric circuits, parameter mirroring, or antithetic sampling.
- These techniques can reduce the total number of shots while maintaining effective convergence, in particular in hardware-constrained settings.

*6.2.9.4 Mini-batching over data*

If the training dataset $\{(x_i, y_i)\}$ is large, one can further reduce variance by averaging gradients over **mini-batches** of size $B$:

$$\widehat{\nabla_\theta}\ \mathscr{L} = \frac{1}{B}\sum_{i=1}^{B} \nabla_\theta\ \mathscr{L}(f(x_i;\ \boldsymbol{\theta}),\ y_i). \tag{6.37}$$

Mini-batching:

- Smooths out per-sample gradient fluctuations,
- Is compatible with shot noise averaging, and
- Allows parallelization across quantum circuits or devices.

However, if each $f(x_i)$ must be measured separately, mini-batching increases the circuit evaluation count. Thus, batch size and shot count must be jointly tuned to balance runtime and statistical robustness.

*6.2.9.5 Practical tips*

- Start with low shot counts (e.g. 100–500), then increase adaptively as training progresses or loss stabilizes.
- Monitor gradient norms and loss plateaus to detect shot-limited training.
- Apply gradient clipping to prevent noisy updates from destabilizing learning.
- Use momentum-based optimizers (e.g., Adam) that average out short-term fluctuations.

### 6.2.10 Loss landscapes and barren plateaus

One of the most fundamental challenges in training variational quantum circuits is the **barren plateau phenomenon**. Barren plateaus refer to regions in the parameter space where the **gradient of the loss function becomes exponentially small**, making training via gradient descent infeasible. Understanding when and why these flat landscapes arise is crucial for designing effective and scalable quantum machine learning models.

### 6.2.11 Intuition behind barren plateaus

In classical optimization, flat regions in the loss landscape slow down training because gradient magnitudes are too small to guide meaningful updates. In quantum circuits, this issue can be much more severe due to the **exponential size of the Hilbert space** [31].

When a variational circuit becomes too expressive, it approximates a 2-design, meaning that it essentially ‘scrambles’ input states uniformly across Hilbert space. As a result:

- The expected gradient of any observable becomes zero.
- The variance of the gradient shrinks exponentially with the number of qubits.

Mathematically, this results in

$$\mathbb{E}\left[\frac{\partial \mathscr{L}}{\partial \theta_i}\right] = 0, \quad \mathrm{Var}\left[\frac{\partial \mathscr{L}}{\partial \theta_i}\right] \in \mathcal{O}\left(\frac{1}{\mathrm{poly}(n)}\right) \tag{6.38}$$

or even worse, in $\mathcal{O}(e^{-n})$, depending on the ansatz and cost function.

#### *6.2.11.1 When do barren plateaus occur?*

Barren plateaus are not universal, they depend on specific circuit choices and cost functions. Key findings include:

- **Global cost functions** (e.g. fidelity between full states) are more susceptible.
- **Deep random circuits** (or those approximating 2-designs) show exponential gradient decay.
- **Unstructured ansatzes** with too much expressivity suffer worse than task-specific circuits.
- **Problem-invariant initializations** (e.g. all-zero angles) often land directly in flat regions.

Recent research has formalized these results and identified several conditions under which barren plateaus arise.

#### *6.2.11.2 Avoiding barren plateaus*

Several strategies have been developed to mitigate this phenomenon:

- **Local cost functions**. Use observables that only depend on a small number of qubits (e.g. local Hamiltonian terms):

$$\mathscr{L} = \langle Z_i Z_{i+1} \rangle. \tag{6.39}$$

  This prevents averaging over the full state space and preserves meaningful gradients [32].
- **Layerwise training**. Gradually grow the circuit depth during training. This keeps the parameter landscape smoother in early stages [33].
- **Problem-inspired ansatzes**. Use architectures derived from the structure of the physical system (e.g. hardware efficient or QAOA circuits) [34, 35].
- **Parameter initialization schemes**. Break symmetry using random initializations or schemes such as identity blocks or shallow pretraining [36].
- **Quantum natural gradient methods**. Incorporate curvature information to traverse plateaus more effectively.

These insights help to co-design circuits and training routines that avoid untrainable regimes, an especially important consideration for scaling to larger quantum systems.

## 6.3 Architectural variants: QCNNs and quantum graph-based models

Variational quantum classifiers can be realized through a variety of circuit architectures, each imposing different structural inductive biases and capabilities. Two prominent classes of models that extend beyond simple layered ansatzes are **quantum convolutional neural networks (QCNNs)** [37] and **quantum graph-based models**, such as **quantum graph neural networks (QGNNs)** [38].

These architectures demonstrate how variational circuits can incorporate spatial locality, data structure symmetries, and hierarchical processing, concepts borrowed from their classical counterparts but adapted to quantum computation.

### 6.3.1 Quantum convolutional neural networks

QCNNs are variational architectures inspired by classical convolutional neural networks (CNNs), which have proven remarkably effective in processing high-dimensional structured data such as images, videos, and spatially organized sensor inputs. QCNNs translate this architectural principle into the quantum domain, where quantum circuits are used to extract hierarchical features from quantum states using local entangling operations and measurement-based pooling. QCNNs are particularly well-suited for tasks involving translation-invariant or lattice-structured data.

A QCNN consists of a layered quantum circuit comprising three key components:

1. **Quantum convolutional layers:** These layers apply parameterized, local unitary operations across subsets of neighboring qubits. The arrangement is translation-invariant, analogous to the application of a filter kernel across an image in a classical CNN. Typically, the convolutional layer includes:
    - Two-qubit or three-qubit entangling gates (e.g. CZ, CNOT, CRY),
    - Followed by single-qubit rotations (e.g. RY, RZ).

    These operations are repeated across the entire register using shared parameters, capturing local correlations and symmetry features.
2. **Pooling layers (quantum downsampling)**: To reduce the number of active qubits and aggregate information, QCNNs implement pooling either through:
    - **Measurement and discard:** Certain qubits are measured, and their outcomes are used to conditionally update the rest (non-unitary, requires classical feedback).
    - **Unitary pooling:** Learned unitaries disentangle and isolate a subset of qubits, which are then traced out or ignored.

    Pooling reduces the circuit width, enabling logarithmic depth scaling, which is favorable for implementation on NISQ hardware.
3. **Fully connected or readout layer:** At the final stage, the remaining few qubits are used to compute the model's output. A final measurement (typically in the $Z$-basis) of a designated readout qubit produces the classification result.

To highlight the unique properties of QCNNs, table 6.3 provides a consolidated summary of their architectural features and implementation strengths. The table outlines how QCNNs leverage hierarchical structures, exploit logarithmic depth scaling, and align well with the symmetry and locality found in many physical systems. Each property is paired with a short explanation to clarify its role within the model architecture and its practical implications for NISQ-era deployment.

**Table 6.3.** Summary of key architectural and implementation advantages of QCNNs.

| Property | Explanation |
|---|---|
| Hierarchical feature extraction | Emulates classical CNNs by capturing local correlations at multiple scales. |
| Logarithmic depth scaling | Pooling operations reduce active qubits, keeping circuits shallow. |
| Noise resilience | Shorter depth and fewer parameters improve robustness to decoherence. |
| Physics-inspired structure | Matches symmetries of many-body systems and translationally invariant data. |
| Hardware alignment | Suited to 1D chains with nearest-neighbor coupling. |

### 6.3.2 Quantum graph neural networks

Quantum graph neural networks (QGNNs) generalize variational quantum circuits to the setting of graph-structured data. Inspired by classical graph neural networks (GNNs), QGNNs aim to capture the topological and relational structure of datasets defined over nodes and edges, such as molecules, social networks, particle interactions, or discretized spatial domains in physics. They key insight behind QGNNs is that quantum entanglement can naturally encode correlations across the edges of a graph, and quantum interference can amplify or suppress global properties emerging from local interactions.

In classical GNNs, message-passing operations iteratively update node features by aggregating information from neighboring nodes. QGNNs seek to reproduce or extend this capability using quantum operations, either by encoding graphs into quantum circuits directly or by embedding node and edge attributes into a quantum state processed by a variational circuit whose structure is influenced by the graph.

There is currently no universally accepted architecture for QGNNs, but most proposals follow one of two broad paradigms:

- **Static QGNNs:** The quantum circuit's layout mirrors the topology of the input graph. Each node corresponds to a qubit or register of qubits, and edges define interactions via entangling gates. This approach is conceptually simple but requires flexible compilation pipelines, in particular for irregular graphs.
- **Dynamic QGNNs:** The graph is first embedded or encoded into a fixed-size quantum state, and a general-purpose variational circuit is used to process this state. Information about the graph structure is introduced through parameter initialization, gate placement, or feature re-uploading. This method decouples circuit topology from input topology but may lose some inductive bias.

A common component of QGNNs is the use of **edge-conditioned gates**, where parameters or even the presence of a gate is modulated by edge weights or features. For instance, an entangling gate $U_{ij}$ between qubits $i$ and $j$ may be turned on only if nodes $i$ and $j$ are connected in the input graph.

Table 6.4 summarizes the architectural principles and implementation benefits associated with QGNNs. These models are explicitly designed to process graph-

**Table 6.4.** Summary of key architectural and implementation advantages of QGNNs.

| Property | Explanation |
|---|---|
| Graph-aware modeling | QGNNs incorporate graph topology directly into circuit design, allowing the learning of relational and structural dependencies between data points. |
| Entanglement along edges | Edge-conditioned entangling gates capture correlations dictated by the input graph, enhancing representational power over classical message passing. |
| Scalability via sparsity | Sparse graphs translate into shallow circuits with fewer entangling operations, enabling more feasible implementation on NISQ devices. |
| Permutation and symmetry invariance | QGNNs can be designed to respect graph symmetries (e.g. node permutation), improving generalization and reducing redundancy in learning. |
| Flexible architecture design | Supports both static (graph-shaped circuits) and dynamic (input-dependent embeddings) models, adaptable to different quantum hardware constraints. |

**Table 6.5.** Comparison of architectural variants of variational quantum classifiers.

| Architecture | Input structure | Strengths | Limitations |
|---|---|---|---|
| QCNN | Grid/lattice data | Hierarchical feature extraction; noise-resilient due to shallow depth | Not well-suited for arbitrary graph data or unstructured inputs |
| QGNN | Arbitrary graphs | Natural for molecular and topological data; models correlations via entangling gates | Mapping to hardware topology can be complex; parameter scaling with graph size |

structured data by embedding topological information into quantum circuits through edge-conditioned entangling gates or graph-aware parameterizations. As the table outlines, QGNNs benefit from inherent scalability when applied to sparse or symmetric graphs, where the circuit depth and parameter count can be kept low. Their flexibility allows for both static and dynamic formulations, depending on the desired trade-off between expressivity and compile-time complexity. Moreover, QGNNs align well with tasks that involve relational reasoning, such as molecular property prediction, where classical GNNs already exhibit strong performance. The ability of QGNNs to encode complex graph correlations in entangled states positions them as promising candidates for quantum-enhanced learning over non-Euclidean data.

To aid the reader in navigating the diversity of circuit-based quantum classifiers, table 6.5 provides a comparative overview of prominent architectural variants. Each model is evaluated based on structural characteristics (such as depth, connectivity, and modularity), trainability considerations, and compatibility with current quantum hardware. This includes QCNNs and QGNNs. The table highlights trade-offs between expressivity and implementation feasibility, helping clarify which architectures are better suited for different data modalities or tasks.

## 6.4 Open challenges and future directions

Despite their theoretical appeal and expressive potential, variational quantum classifiers and neural networks remain difficult to scale and deploy in practice. This section outlines some of the core challenges faced in this domain and identifies several promising opportunities for future research and hardware-software co-design.

### 6.4.1 Trainability and optimization

One of the most pressing challenges in training variational classifiers is the emergence of **barren plateaus**, regions in parameter space where the gradient vanishes exponentially with system size. These regions hinder the optimizer's ability to explore the landscape effectively and result in slow or stalled convergence. The problem is particularly acute in deep circuits, randomly initialized ansatzes, or global cost functions. While initialization strategies, local cost functions, and symmetry-preserving architectures (such as QCNNs or QGNNs) offer partial mitigation, a systematic theory of trainability remains an open research frontier.

Moreover, noisy gradient estimation due to finite sampling introduces significant variance during training, in particular when computing gradients via the parameter-shift rule. This effect compounds with the circuit depth and number of parameters, making training on real hardware both time-consuming and error-prone. Recent advances in adaptive batching, momentum-aware optimizers, and gradient variance reduction techniques show promise, but their integration with quantum circuits is still under active development.

### 6.4.2 Resource overhead and compilation

Variational models often require deep circuits or repeated evaluations of the quantum state, leading to substantial resource overhead on NISQ hardware. For example, encoding high-dimensional input data may require redundant qubit registers or repeated feature re-uploading layers. Furthermore, compilation of problem-specific architectures, such as those used in QGNNs, introduces additional complexity, in particular when circuit layout must reflect irregular input structures such as sparse graphs or molecules.

Opportunities lie in **hardware-aware ansatz design**, where the circuit structure is tailored to the physical qubit layout, and parameter-efficient encoding strategies that minimize qubit count without sacrificing expressivity. Leveraging tensor network-inspired architectures, or exploiting quantum hardware primitives such as mid-circuit measurements and feedback, could help circumvent current resource bottlenecks.

### 6.4.3 Generalization and inductive bias

Unlike classical deep learning models, quantum classifiers still lack a well-developed theory of generalization. Empirical results suggest that quantum circuits can fit training data with high fidelity, but their ability to extrapolate to unseen examples remains poorly understood [39–41]. Furthermore, because quantum models do not always rely on compositional or translation-invariant representations, encoding appropriate inductive biases, in particular for structured data, is crucial.

There is a rich opportunity to develop hybrid training protocols that combine classical neural architectures (e.g. convolutional or attention layers) with quantum subroutines acting as non-linear kernels or feature transformers. These schemes can introduce structure into quantum models while still leveraging their non-classical expressivity. Similarly, symmetry-informed architectures, such as equivariant QGNNs, may improve generalization by enforcing desirable invariances directly into the circuit.

### 6.4.4 Benchmarking and expressivity analysis

A persistent bottleneck in the field is the lack of robust benchmarking datasets and evaluation criteria tailored to hybrid and variational quantum models. Most demonstrations still rely on synthetic tasks or low-dimensional datasets that are easily solvable classically. While these are necessary for proof-of-concept, they limit our ability to assess quantum advantage in real-world conditions.

Addressing this gap requires the creation of **domain-relevant benchmarks**, in particular from fields such as chemistry, materials science, and high energy physics. Additionally, the development of expressivity metrics, such as entangling power, effective dimension, or Fisher information, can help quantify the representational capacity of variational models and guide ansatz design.

### 6.4.5 Opportunity in co-design

Perhaps the most exciting frontier lies in **co-design quantum models and hardware architectures simultaneously**. As quantum hardware continues to evolve, the opportunity to define models that are both *hardware-efficient* and *algorithmically powerful* becomes increasingly feasible. In this sense, models such as QCNNs and QGNNs are not just convenient approximations but are tailor-made for the constraints and symmetries of NISQ devices.

Moreover, the integration of domain knowledge into variational architectures could enable problem-specific quantum advantage, where quantum models outperform classical ones not in general, but for strategically chosen, high-value tasks.

## 6.5 Summary

In this chapter, we explored the landscape of variational quantum classifiers and their integration into hybrid quantum–classical neural network models. Building on the variational optimization framework introduced in the previous chapter, we presented how parameterized quantum circuits can be trained as learning models for supervised classification tasks. Beginning with an overview of classical neural networks and their core mathematical structure, we introduced their quantum counterparts and examined how they can be embedded into hybrid workflows that alternate between classical and quantum computation.

We then focused on the techniques that enable the training of such hybrid models, including gradient-based optimization and the implementation of backpropagation through parameter-shift rules and adjoint differentiation. Particular attention was given to the sources of gradient variance, such as shot noise and stochastic batching, and their implications for convergence and robustness in practical training loops.

Architectural variants, including QCNNs and QGNNs, were analysed in terms of their inductive biases, depth efficiency, and suitability for structured data. The chapter concluded with a discussion of core challenges as well as opportunities afforded by co-design strategies, expressivity theory, and domain-specific quantum model architectures. These discussions lay the groundwork for understanding quantum neural networks not only as theoretical constructs but as practical tools in near-term quantum machine learning.

## References

[1] McCulloch W S and Pitts W 1943 A logical calculus of the ideas immanent in nervous activity *Bull. Math. Biophys.* **5** 115–33

[2] Altaisky M V 2001 Quantum neural network arXiv: quant-ph/0107012

[3] Schuld M, Sinayskiy I and Petruccione F 2015 Simulating a perceptron on a quantum computer *Phys. Lett.* A **379** 660–3

[4] Kak S 1995 On quantum neural computing *Inf. Sci.* **83** 143–60

[5] Pérez-Salinas A, Cervera-Lierta A, Gil-Fuster E and Latorre J I 2020 Data re-uploading for a universal quantum classifier *Quantum* **4** 226

[6] Cybenko G 1989 Approximation by superpositions of a sigmoidal function *Math. Control Signals Syst.* **2** 303–14

[7] Hornik K 1991 Approximation capabilities of multilayer feedforward networks *Neural Networks* **4** 251–7

[8] Schäfer A M and Zimmermann H G 2006 Recurrent neural networks are universal approximators *Artificial Neural Networks—ICANN 2006* Lecture Notes in Computer Science vol 4131 *(Berlin: Springer) pp 632–40*

[9] Grant E, Benedetti M, Cao S, Hallam A, Lockhart J, Stojevic V, Green A G and Severini S 2018 Hierarchical quantum classifiers *npj Quantum Inf.* **4** 65

[10] Chen H, Wossnig L, Severini S, Neven H and Mohseni M 2020 Universal discriminative quantum neural networks *Quant. Mach. Intell.* **3** 1

[11] Mitarai K, Negoro M, Kitagawa M and Fujii K 2018 Quantum circuit learning *Phys. Rev.* A **98** 032309

[12] Cong I, Choi S and Lukin M D 2019 Quantum convolutional neural networks *Nat. Phys.* **15** 1273–8

[13] Wan K H, Dahlsten O, Kristjánsson H, Gardner R and Kim M S 2017 Quantum generalisation of feedforward neural networks *npj Quantum Inf.* **3** 36

[14] Baydin A G, Pearlmutter B A, Radul A A and Siskind J M 2018 Automatic differentiation in machine learning: a survey *J. Mach. Learn. Res.* **18** 1–43

[15] Paszke A *et al* 2019 Pytorch: an imperative style, high-performance deep learning library arXiv: 1912.01703
[16] Abadi M *et al* 2016 Tensorflow: a system for large-scale machine learning *Proc. 12th USENIX Conf. on Operating Systems Design and Implementation* pp 265–83
[17] Lin M 2024 Automatic functional differentiation in JAX arXiv: 2311.18727
[18] Bergholm V *et al* 2022 PennyLane: automatic differentiation of hybrid quantum-classical computations arXiv: 1811.04968
[19] Emre Sahin M, Altamura E, Wallis O, Wood S P, Dekusar A, Millar D A, Imamichi T, Matsuo A and Mensa S 2025 Qiskit Machine Learning: an open-source library for quantum machine learning tasks at scale on quantum hardware and classical simulators arXiv: 2505.17756
[20] Broughton M *et al* 2021 Tensorflow quantum: a software framework for quantum machine learning arXiv: 2003.02989
[21] Schuld M and Killoran N 2019 Quantum machine learning in feature Hilbert spaces *Phys. Rev. Lett.* **122** 040504
[22] Wierichs D, Izaac J, Wang C and Yen-Yu Lin C 2022 General parameter-shift rules for quantum gradients *Quantum* **6** 677
[23] Spall J C 1992 Multivariate stochastic approximation using a simultaneous perturbation gradient approximation *IEEE Trans. Automatic Control* **37** 332–41
[24] Jones T and Gacon J 2020 Efficient calculation of gradients in classical simulations of variational quantum algorithms arXiv: 2009.02823
[25] Stokes J, Izaac J, Killoran N and Carleo G 2020 Quantum natural gradient *Quantum* **4** 269
[26] Braunstein S L and Caves C M 1994 Statistical distance and the geometry of quantum states *Phys. Rev. Lett.* **72** 3439–43
[27] Braunstein S L, Caves C M and Milburn G J 1996 Generalized uncertainty relations: theory, examples, and Lorentz invariance *Ann. Phys.* **247** 135–73
[28] Meyer J J, Borregaard J and Eisert J 2021 A variational toolbox for quantum multi-parameter estimation *npj Quantum Inf.* **7** 89
[29] Gu A, Lowe A, Dub P A, Coles P J and Arrasmith A 2021 Adaptive shot allocation for fast convergence in variational quantum algorithms arXiv: 2108.10434
[30] Kübler J M, Arrasmith A, Cincio L and Coles P J 2020 An adaptive optimizer for measurement-frugal variational algorithms *Quantum* **4** 263
[31] McClean J R, Boixo S, Smelyanskiy V N, Babbush R and Neven H 2018 Barren plateaus in quantum neural network training landscapes *Nat. Commun.* **9** 4812
[32] Cerezo M, Sone A, Volkoff T, Cincio L and Coles P J 2021 Cost function dependent barren plateaus in shallow parametrized quantum circuits *Nat. Commun.* **12** 1791
[33] Skolik A, McClean J R, Mohseni M, van der Smagt P and Leib M 2021 Layerwise learning for quantum neural networks *Quant. Mach. Intell.* **3** 5
[34] Park C-Y, Kang M and Huh J 2024 Hardware-efficient ansatz without barren plateaus in any depth arXiv: 2403.04844
[35] Larocca M, Czarnik P, Sharma K, Muraleedharan G, Coles P J and Cerezo M 2022 Diagnosing barren plateaus with tools from quantum optimal control *Quantum* **6** 824
[36] Grant E, Wossnig L, Ostaszewski M and Benedetti M 2019 An initialization strategy for addressing barren plateaus in parametrized quantum circuits *Quantum* **3** 214
[37] Cong I, Choi S and Lukin M D 2019 Quantum convolutional neural networks *Nat. Phys.* **15** 1273–8

[38] Verdon G, McCourt T, Luzhnica E, Singh V, Leichenauer S and Hidary J 2019 Quantum graph neural networks arXiv: 1909.12264

[39] Caro M C, Huang H-Y, Cerezo M, Sharma K, Sornborger A, Cincio L and Coles P J 2022 Generalization in quantum machine learning from few training data *Nat. Commun.* **13** 4919

[40] Wang Y and Qi B 2023 Enhanced generalization of variational quantum learning under reduced-domain initialization *2023 42nd Chin. Control Conf. (CCC)* pp 6771–6

[41] Banchi L, Pereira J and Pirandola S 2021 Generalization in quantum machine learning: a quantum information standpoint *PRX Quantum* **2** 040321

**IOP** Publishing

Quantum Machine Learning
Concepts and possibilities
**Andrea Delgado and Kathleen E Hamilton**

# Chapter 7

# Quantum generative models

Generative models play a foundational role in machine learning by enabling systems to learn and replicate the underlying structure of data distributions. Unlike discriminative models (discussed in chapter 6), which aim to distinguish between different classes or outputs, generative models are useful for unconditioned density estimation given a set of unlabeled data features. This capability allows them to synthesize new data samples, estimate probabilities, and support a wide range of applications, including image generation, data augmentation, anomaly detection, and unsupervised representation learning.

Quantum generative models (QGMs) extend these capabilities into the quantum domain, leveraging quantum correlations to learn and represent complex distributions that may be intractable for classical models. Unitary quantum dynamics can implicitly represent probability amplitudes without the additional overhead associated with evaluating partition functions. Leveraging quantum dynamics and the expressivity of quantum systems may lead to richer families of distributions, or compact representations of data, or powerful candidates for sampling tasks that underlie generative modeling. The prospect of quantum advantage, demonstrating that quantum models can outperform classical ones in practical, real-world tasks, is a compelling motivation for studying QGMs. However, a stronger motivation is also found in the close connection of quantum dynamics and probabilistic model structure.

In recent years, many QGM frameworks have been proposed, ranging from energy-based models such as quantum Boltzmann machines to circuit-based architectures such as quantum circuit Born machines and quantum generative adversarial networks. This chapter explores the theory and practice of QGMs. We begin by reviewing classical probabilistic models and generative models, paying special attention to the Boltzmann machine [1–3] which was one of the early adapted quantum models. We then introduce several classes of quantum generative models, discuss their training methods, and examine the concept of expressivity, which is

doi:10.1088/978-0-7503-4952-9ch7

critical to understanding the power and limitations of these models. Finally, we highlight current challenges, and open research directions for the development and deployment of quantum generative models.

Throughout this chapter, we aim to balance conceptual clarity and technical rigor to provide the tools needed to understand, evaluate, and implement generative models in quantum machine learning pipelines. Whether used for modeling quantum data or enhancing classical tasks, QGMs offer a promising frontier in the effort to harness quantum computing for learning tasks.

## 7.1 Classical generative models

Classical generative models $\mathcal{G}$ are probabilistic models that can be parametric or non-parametric. They are used to represent the probability distribution of a dataset ($\mathcal{D}(\mathbf{x})$) that is comprised of observations ($\mathbf{x}$) and may or may not include labels ($y$). Generative models learn joint distributions $p(x, y)$ or the marginal $p(x)$,and trained models can be used to generate new samples that resemble the original data, estimate likelihoods, or infer missing or hidden variables making them ideal for tasks involving synthesis, completion, and unsupervised learning.

Parametric models are defined by a finite number of parameters and assume that observations are generated according to a specific family of distributions. For example, **Gaussian mixture models (GMMs)**, assume that observations are drawn from a process that can be represented by the weighted combination of several base distributions which are assumed to be Gaussian with unknown parameters (weights, means, and standard deviations). Mixture models can be constructed using other distributions as well. Another example, **hidden Markov models (HMMs)**, assume that observations $\mathbf{x}$ are generated sequentially, or by assuming that the system follows a Markov process in the hidden states.

These models can be generalized under **probabilistic Graph models** [4]. Optimizing such models often employs techniques such as expectation-maximization [5] or belief propagation [6]. These models are interpretable and computationally tractable for low-dimensional or structured problems, but they struggle with high-dimensional, complex datasets due to their limited representational capacity.

Non-parametric models rely on an unbounded number of parameters. This covers energy-based models, Bayesian models, and state of the art deep learning models that utilize neural networks.

### 7.1.1 Boltzmann machines and energy-based models

Boltzmann machines (BMs) [1], their restricted variants (RBMs) [7], and the generalization to deep belief networks (DBNs) [8] are generative models where data samples correspond to low-energy configurations of a system described by a global energy function. Structurally the models are neural networks where neurons are partitioned into visible ($\mathbf{v}$) and hidden variables ($\mathbf{h}$). The BM is a fully connected, recurrent network, whereas the RBM and DBN restrict weighted connections into bipartite graphs (no direct connections exist between visible variables or hidden variables).

A classical Boltzmann machine defines a probability distribution over the visible variables by associating each configuration $\mathbf{x}$ (comprising visible and hidden units) with an energy value. The model learns by minimizing the energy of likely configurations and maximizing the energy of unlikely ones. The probability of a visible configuration $\mathbf{v}$ is given by marginalizing over hidden variables $\mathbf{h}$,

$$p(\mathbf{v}) = \frac{1}{Z}\sum_{\mathbf{h}} e^{-E(\mathbf{v},\,\mathbf{h})}, \tag{7.1}$$

where the energy function is defined as

$$E(\mathbf{v}, \mathbf{h}) = -\mathbf{v}^{\top} W\mathbf{h} - \mathbf{b}^{\top}\mathbf{v} - \mathbf{c}^{\top}\mathbf{h}, \tag{7.2}$$

and the partition function $Z$ ensures normalization:

$$Z = \sum_{\mathbf{v},\mathbf{h}} e^{-E(\mathbf{v},\,\mathbf{h})}. \tag{7.3}$$

The summation over all pairs of visible and hidden units in the partition function makes training BMs difficult. The contrastive divergence algorithm [9] has been successful in training RBMs. Since no intra-layer connections are allowed, the states of the hidden or visible neurons are iteratively updated and stochastic sampling offers a tractable alternative that became widely used in deep learning pipelines, in particular as pre-training layers for DBNs.

BMs are conceptually appealing because they encode correlations through learned energy landscapes and can be related to physical systems in thermal equilibrium. This connection makes them ideal candidates for generalization to the quantum domain. The energy-based nature of these models makes them particularly appealing for statistical physics-inspired learning and sets the stage for quantum extensions that substitute classical energy functions with quantum Hamiltonians.

RBMs gained popularity in the early 2000s as building blocks of DBNs and were used for dimensionality reduction, collaborative filtering, and pre-training deep networks. After the advent of deep learning, energy-based models [10] have incorporated other deep neural architectures into non-parametric models trained by unsupervised learning.

### 7.1.2 Deep generative models

Modern deep generative models have greatly expanded the applicability of generative modeling to high-dimensional data, including images, text, and audio. Notable examples include:

- **Variational autoencoders (VAEs)** [11]: Learn a probabilistic latent space using an encoder–decoder architecture, optimizing a variational lower bound on the data likelihood.
- **Generative adversarial networks (GANs)** [12]: Consist of a generator and a discriminator trained in a minimax game, leading to highly realistic sample generation.

**Table 7.1.** Classical generative models at a glance.

| Model Type | Example | Key Mechanism | Strengths | Limitations |
|---|---|---|---|---|
| Probabilistic models | GMM, HMM | Parametric estimation | Interpretable, well-studied | Limited in high dimensions |
| Energy-based models | RBM | Low-energy sampling | Physically inspired, flexible | Intractable partition function |
| Latent variable models | VAE | Variational inference | Probabilistic decoding | Blurry samples |
| Adversarial models | GAN | Generator–discriminator game | Sharp, realistic outputs | Training instability, mode collapse |
| Invertible models | Normalizing flow | Bijective mapping from prior | Exact likelihoods | Requires specific architectural design |

- **Normalizing flows** [13]: Learn invertible transformations from simple distributions (e.g. Gaussians) to complex ones, enabling exact likelihood computation.

These models offer powerful tools for generation, representation learning, and simulation, but also come with limitations (table 7.1).

## 7.2 Quantum generative models

Quantum generative models (QGMs) have been developed either as hardware-agnostic theoretical models or have been developed through co-design with near-term quantum hardware. Many technologies are used for quantum processors and different processors rely on different computing models. Quantum annealers use the adiabatic evolution of quantum systems [14], while digital quantum computers use unitary evolution of quantum systems. Annealers are well-suited for energy-based models, while digital quantum systems are well-suited for variational quantum algorithms with parameterized quantum circuit models. However, hardware capabilities are advancing yearly and are assisted through novel algorithm design.

A QGM is typically defined by three components:

- **A quantum circuit** $U(\theta)$ with parameters $\boldsymbol{\theta}$, acting on an initial state $|\Psi\rangle$;
- **A measurement scheme**, often in the computational basis; and
- **A training objective**, such as matching the output distribution $p_\theta(\mathbf{x})$ to a target distribution $p_{\text{data}}(\mathbf{x})$ via a loss function $\mathcal{L}(p_\theta, p_{\text{data}})$.

The output distribution is defined as

$$p_\theta(\mathbf{x}) = |\langle \mathbf{x}| U(\boldsymbol{\theta})|0^{\otimes n}\rangle|^2. \tag{7.4}$$

Training proceeds by adjusting $\boldsymbol{\theta}$ to minimize a divergence or distance metric, total variation distance, or adversarial loss. Depending on the task, QGMs can be used for:

- **Quantum data generation**, i.e. reproducing quantum states or measurement statistics;
- **Classical data modeling**, i.e. approximating the distribution of classical data using a quantum circuit; and
- **Hybrid applications**, where classical and quantum components interact during training or inference.

### 7.2.1 Quantum Boltzmann machines

The motivation for Boltzmann machines, and many neural network architectures, is the emulation of neural units in cognition. When represented as a network of stochastic, binary variables with iteractions mitigated through weighted connections, the model is closely related to the Ising spin glass model. These models laid the foundation for many developments in unsupervised learning and probabilistic modeling and have since influenced a variety of quantum generative architectures.

Quantum Boltzmann machines (QBMs) [15] extend classical BMs by replacing classical energy functions with Ising spin Hamiltonians. Instead of modeling probability distributions directly, QBMs prepare a thermal state in a Hilbert space. Through projective measurements, samples can be drawn according to the density matrix of the prepared state:

$$\rho = \frac{e^{-\beta H}}{\mathrm{Tr}(e^{-\beta H})} = \sum\nolimits_i p_i |\psi_i\rangle\langle\psi_i|. \tag{7.5}$$

Here, $H$ is a Hamiltonian encoding interactions between qubits, $\beta$ is the inverse temperature $(1/k_{\mathrm{B}}T)$,[1] and $|\psi_i\rangle$ represents individual state vectors. The transverse field Ising model is a well-established system used for Boltzmann machines the Hamiltonian is composed of nearest-neighbor interactions as well as on-site potentials and external fields,

$$H_{\mathrm{ISING}} = -\sum\nolimits_{\langle i,j\rangle} J_{ij}\sigma_i\sigma_j - \mu\sum\nolimits_i h_i\sigma_i. \tag{7.6}$$

This system exhibits phase transitions where individual spins (i.e. neurons or binary variables) can align or anti-align with nearest neighbors, or can algin or anti-align with the external field. These interactions are controlled by the amplitude and sign of the individual terms $J_{ij}$, and the on-site potential and external field terms $(h_i)$ and can result in long-range coherent patterns or random, frustrated configurations of spins (variables).

Training this model approximates a target distribution by optimizing the parameters $(J_{ij}, h_i)$ of $H_{\mathrm{ISING}}$. The probability of a classical configuration $\mathbf{v}$ is given by

[1] $k_{\mathrm{B}}$ is the Boltzmann constant.

$$p(\mathbf{v}) = \langle \mathbf{v}|\rho|\mathbf{v}\rangle = \frac{\langle \mathbf{v}|e^{-\beta H}|\mathbf{v}\rangle}{\mathrm{Tr}(e^{-\beta H})}. \tag{7.7}$$

While conceptually powerful, many challenges for QBMs exist. Methods for efficient thermal state preparation methods are needed to ensure that candidate states can be prepared and are stable enough so that sufficient samples can be drawn on noisy hardware. Alternatively, when generating samples with classical simulation or emulation, efficient Monte Carlo methods are needed. Finally, understanding the optimization landscape is necessary to build efficient training methods that can converge to good solutions under noise, both hardware and stochastic. One example is the use of quantum annealing to navigate rugged optimization landscapes [16].

Nevertheless, QBMs build genuinely quantum generative models, using the formalism of quantum statistical mechanics to represent and learn complex distributions. Table 7.2 summarizes the conceptual and operational differences between RBMs and their quantum counterparts, QBMs. While RBMs operate with classical binary variables and rely on energy functions defined over visible and hidden nodes, QBMs replace these with quantum states governed by a Hamiltonian and described via thermal density matrices. The table contrasts their respective sampling strategies, expressive capacities, and practical training challenges, highlighting how quantum models may potentially overcome some of the limitations inherent to classical energy-based approaches. Notably, QBMs are able to exploit quantum coherence and entanglement to model more complex correlations, although this comes at the cost of increased implementation complexity and hardware requirements.

Beyond the framework of energy-based models, QGMs use quantum mechanical properties of superposition, entanglement, and interference to model probability distributions. With parameterized unitary gates, QGMs model generative processes that either sample from, or approximate, complex data distributions. These models

**Table 7.2.** Comparison of RBMs and QBMs. QBMs generalize RBMs by replacing classical energy functions with quantum Hamiltonians, allowing for enhanced representational power via quantum effects such as superposition and entanglement.

| Feature | Classical RBM | Quantum Boltzmann machine |
|---|---|---|
| Energy function | $E(\mathbf{v}, \mathbf{h})$ | Quantum Hamiltonian $H$ |
| State representation | Binary vector | Quantum density matrix $\rho$ |
| Sampling | Gibbs sampling/contrastive divergence | Quantum thermal state sampling |
| Expressivity | Limited by architecture | Enhanced by quantum coherence and entanglement |
| Training challenges | Approximate gradient estimation | Intractable partition function, hardware required |

can be trained to match classical data distributions, generate quantum states, or even learn hybrid distributions across classical and quantum domains.

In the following subsections, we outline the general non-parametric framework of quantum generative modeling and introduce some of the most widely studied QGM architectures, organized by their structure and target output. These models are built on **parameterized quantum circuits (PQCs)**, and the representation and expressivity is dependent on the number of trainable parameters. PQCs are composed of single-qubit rotation gates and entangling gates arranged in layers or repeating motifs, and are trained via classical optimization loops to approximate a target distribution. We now explore the main model architectures in detail.

### 7.2.2 Quantum circuit Born machines

Quantum circuit Born machines (QCBMs) [17–19] generate samples according to the squared amplitudes of a prepared quantum state. This is the well-known **Born rule** of quantum mechanics and the parameterized probability distribution is defined as

$$p_{\theta}(\mathbf{x}) = |\langle \mathbf{x}| U(\boldsymbol{\theta})|0^{\otimes n}\rangle|^2. \tag{7.8}$$

The PQC $U(\theta)$ is constructed with repeating patterns of parameterized single-axis rotations ($R_X(\theta)$, $R_Y(\theta)$, $R_Z(\phi)$), or using general Euler rotations $U_3(\theta, \phi, \lambda)$ and **entangling gates**, multi-qubit gates. Commonly PQCs are designed to utilize hardware-native gates (e.g. two-qubit gates CNOT or CZ over multi-control or multi-qubit unitaries) and rely on interactions that match the sparse connections found on near-term hardware (e.g. linear or circular nearest-neighbor connectivity).

We distinguish QGMs from the energy-based models by the fact that the circuit design is not driven by an underlying system Hamiltonian. During training, instead of minimizing an energy function, the QCBM is trained by minimizing a loss function $\mathscr{L}(p_{\text{data}}, p_\theta)$. This is evaluated between two distributions: the target distribution $p_{\text{data}}$ and the distribution constructed with samples drawn from the QCBM ($p_\theta$).

There are many loss functions that can be used for training. First, the loss can compare sample sets drawn from the target and the QCBM and assess similarity through statistical tests. As an example, the maximum mean discrepancy (MMD) [20] loss uses a kernel function to compare the similarity between multi-dimensional samples. A second class of loss functions are derived from divergence families[21]. Bregman, Renyi, or general $f$ divergences contain the Kullback–Leibler (KL) divergence, the symmetrized Jensen–Shannon (JS) divergence, the Hellinger distance and the total variation distance (TVD).

**Training is hybrid** after sampling from the quantum circuit to estimate $p_\theta(\mathbf{x})$, a classical optimizer updated the parameters $\boldsymbol{\theta}$ based on the chosen loss. For example, to model the parity distribution over three bits, where only bitstrings with even parity have non-zero probability, the QCBM is trained until

$$p_\theta(000) \approx p_\theta(011) \approx p_\theta(101) \approx p_\theta(110) > 0 \quad \textit{and} \quad p_\theta(\text{oddparity}) \approx 0. \tag{7.9}$$

The expressivity of the circuit (number of layers and qubits) determines which distributions can be learned.

### 7.2.3 Quantum generative adversarial networks

Quantum generative adversarial networks (QGANs) [22] adapt the classical GAN framework [12] into the quantum setting, with a **quantum generator** and a **discriminator** that may be either classical or quantum. The goal is for the generator to produce quantum or classical states indistinguishable from the target data distribution.

The generator prepares a parameterized quantum state:

$$|\psi_G(\boldsymbol{\theta})\rangle = U_G(\boldsymbol{\theta})|0\rangle^{\otimes n} \tag{7.10}$$

which is either:

- Measured to generate classical bitstrings, or
- Kept as a quantum state for comparison with quantum data.

The discriminator distinguishes between real and generated data. In a fully quantum QGAN, it is a parameterized observable or variational circuit $U_D(\phi)$ followed by a measurement. The adversarial objective is

$$\min_{\theta}\max_{\phi} \mathbb{E}_{\mathbf{x}\sim p_{\text{data}}}[D(\mathbf{x};\, \phi)] - \mathbb{E}_{\mathbf{x}\sim p_G}[D(\mathbf{x};\, \phi)]. \tag{7.11}$$

Training alternates between improving the generator to 'fool' the discriminator and updating the discriminator to better distinguish the sources.

### 7.2.4 Quantum variational autoencoders

Quantum variational autoencoders (QVAEs) [23] extend classical VAEs by using quantum circuits in the encoder and/or decoder networks. The basic idea is to map classical inputs to a distribution over quantum latent variables and then decode this to reconstruct or generate new data.

- **Encoder:** Maps data $x$ to a parameterized quantum state $|z(\phi_x)\rangle$, possibly by encoding $x$ into gate angles.
- **Decoder:** A parameterized quantum circuit $U_D(\theta)$ maps the latent state back to a distribution over outputs.

The training objective is the **evidence lower bound (ELBO)**:

$$\mathbb{L}_{\text{ELBO}} = \mathbb{E}_{q(z|x)}[\log p(x|z)] - \text{KL}[q(z|x)\ ||\ p(z)], \tag{7.12}$$

where $q(z|x)$ and $p(z)$ may be encoded by quantum circuits, and evaluation often requires classical post-processing.

## 7.3 Expressivity and learning power

One of the key motivations for using quantum circuits in generative modeling is their **expressivity**, the ability to represent complex probability distributions or quantum states compactly and accurately. Expressivity in quantum generative models is closely tied to the structure of the circuit, the dimensionality of the Hilbert space, and the entanglement and interference effects that quantum operations can exploit.

### 7.3.1 Defining expressivity

The expressivity of a quantum generative model refers to it ability to represent a wide and complex class of probability distributions over bitstrings or other measurement outcomes. In the context of circuit-based models such as QCBMs, this notion is formalized via the **induced distribution** [24, 25], that is, the probability distribution over measurement outcomes defined by the Born rule applied to a parameterized quantum state.

Let $U(\theta)$ be a parameterized quantum circuit acting on an $n$-qubit system initialized in the computational basis state $|0\rangle^{\otimes n}$. The output probability distribution over bitstrings $\mathbf{x} \in \{0, 1\}^n$ is given by

$$p_\theta(\mathbf{x}) = |\langle \mathbf{x}| U(\boldsymbol{\theta})|0^{\otimes n}\rangle|^2. \tag{7.13}$$

The **induced distribution** $p_\theta$ defines the model's output. The set of all distributions accessible by varying the parameters $\boldsymbol{\theta}$ determines the expressive capacity of the model. In contrast to classical neural networks that explicitly compute an output function, QCBMs implicitly define a probability distribution through wavefunction amplitudes and the interference patterns generated by the circuit.

As shown by Benedetti *et al* [17], the expressivity of QCBMs is directly linked to the structure and depth of the circuit ansatz. Shallow circuits or those with restricted connectivity are generally only capable of representing a narrow family of distributions, often those with limited or local correlations. Deeper or more entangling architectures, however, can give rise to highly nontrivial distributions, many of which may be hard to simulate or represent classically.

Sim *et al* [26] introduced a quantitative notion of **expressibility** based on how uniformly the circuit ensemble covers the space of unitary operations, drawing connections to the theory of unitary $t$-designs. Du *et al* [27] further formalized the comparison between classical and quantum models, showing that certain families of quantum circuits can represent probability distributions that require exponentially large classical models to approximate.

Taken together, these results suggest that quantum circuits offer a natural and powerful inductive bias for generative modeling. The expressivity of a QCBM is not only a function of its parameter count or Hilbert space size, but also of its ability to generate entanglement, exploit interference, and produce classically intractable correlations.

### 7.3.2 Factors affecting expressivity

The expressivity of a QGM depends on multiple aspects:

- **Circuit depth and gate set:** Deeper circuits with a universal gate set can approximate arbitrary unitaries.

- **Entanglement structure:** Models that generate highly entangled states tend to have more expressive power.
- **Parameterization:** Redundant or untrainable parameters can reduce effective expressivity.
- **Measurement scheme:** The choice of measurement basis can influence what features are observable in the output.

A useful metric is the **frame potential** [28, 29], which measures how well a set of circuits covers the unitary group. A circuit family with low frame potential is closer to forming a unitary 2-design, which is often used as a proxy for high expressivity.

### 7.3.3 Trade-off with trainability

High expressivity can be a double-edged sword. While expressive models are capable of learning complex distributions, they may also be prone to:

- **Overfitting** in small datasets,
- **Barren plateaus**, where gradients vanish exponentially with system size [30, 31], and
- **Optimization instability** when too many parameters interact nonlinearly.

Thus, balancing expressivity with **inductive bias** and **training feasibility** is critical when designing quantum generative circuits.

## 7.4 Training quantum generative models

Training quantum generative models involves optimizing the parameters of a quantum circuit such as the distribution induced by measurement outcomes approximates a target distribution. While conceptually analogous to training classical generative models, the quantum setting introduces distinctive challenges and opportunities, particularly due to measurement stochasticity, non-differentiability, and hardware constraints.

This section outlines the typical training loop, common loss functions, gradient estimation techniques, and specific difficulties encountered during optimization.

### 7.4.1 The optimization loop

Most quantum generative models follow a hybrid quantum–classical training loop:

- **Initialize parameters $\boldsymbol{\theta}$** for the quantum circuit $U(\theta)$.
- **Generate samples** from the circuit by measuring in the computational basis, yielding outcomes $\mathbf{x} \sim p_\theta(\mathbf{x})$.
- **Estimate a loss function** comparing $p_\theta$ to the target distribution $p_{\text{data}}$.
- **Update parameters** using a classical optimizer (e.g. gradient descent, Adam, or SPSA).
- Repeat until convergence.

This loop is model-agnostic and applies to QCBMs, QGANs, QVAEs, and other quantum generative frameworks.

### 7.4.2 Loss functions

The loss functions guides the model toward better approximation of the target distribution. Common choices include:

- **KL divergence** [32]:

$$D_{\mathrm{KL}}(p_{\text{data}}\,||p_\theta) = \sum_{\mathbf{x}} p_{\text{data}}(\mathbf{x})\log \frac{p_{\text{data}}(\mathbf{x})}{p_\theta(\mathbf{x})}. \tag{7.14}$$

  Requires knowledge of both distributions and can be unstable if $p_\theta(\mathbf{x})$ vanishes.

- **Maximum mean discrepancy (MMD)** [33]:

$$\mathbb{L}_{\mathrm{MMD}} = \mathbb{E}_{x,x'\sim p_{\text{data}}}[k(x,\, x')] + \mathbb{E}_{y,y'\sim p_\theta}[k(y,\, y')] - 2\mathbb{E}_{x\sim p_{\text{data}},y\sim p_\theta}[k(x,\, y)], \tag{7.15}$$

  where $x,\, x' \sim p_{\text{data}},\, y,\, y' \sim p_\theta$, and $k$ is a kernel function (e.g. Gaussian). MMD is popular in QCBMs due to its sample-based and gradient-free nature.

- **Adversarial loss:** Used in QGANs, where a discriminator is trained simultaneously to distinguish real versus generated data.
- **Reconstruction + regularization losses:** Used in QVAEs, combining negative log-likelihood with KL divergence on latent variables.

### 7.4.3 Gradient estimation

Unlike classical neural networks, gradients in quantum models are not obtained via automatic differentiation. Instead, **parameter-shift rules** and **finite difference estimators** are used.

For a parameter $\theta$ appearing in a gate of the form $e^{-i\theta G}$, where $G$ has two distinct eigenvalues $\pm\lambda$, the gradient of an expectation value $\langle O\rangle$ can be computed as

$$\frac{\partial\langle O\rangle}{\partial\theta} = \frac{1}{2\lambda}\left(\langle O\rangle_{\theta+\frac{\pi}{2\lambda}} - \langle O\rangle_{\theta-\frac{\pi}{2\lambda}}\right). \tag{7.16}$$

This allows gradient computation using only circuit evaluations at shifted parameter values.

For non-differentiable objectives (e.g. discrete outputs, sample-based losses such as MMD), gradient-free optimizers are often used. These are robust to noise but may require more iterations to converge.

### 7.4.4 Training challenges

QGMs face similar obstacles during training that are encountered by other variational algorithms:

- **Shot noise and statistical error:** Estimates of loss and gradients rely on finite sampling from quantum hardware, introducing variance.
- **Barren plateaus:** Variational circuits can exhibit exponentially vanishing gradients in high-dimensional Hilbert spaces, in particular with random or deep ansatzes.

- **Mode collapse:** Especially relevant in adversarial training, where the generator fails to capture the diversity of the data distribution. This can also be encountered due to the prevalance of spurious minima in the optimization landscape.
- **Hardware constraints:** Real quantum devices have limited qubit counts, gate fidelity, and coherence times, which restrict circuit complexity and batch sizes.

## 7.5 Case studies and applications

QGMs offer a flexible and powerful framework for learning and synthesizing distributions that are difficult or inefficient to represent classically. This has opened the door to a wide range of applications spanning quantum physics, machine learning, optimization, and sensing. In this section, we highlight some use cases where QGMs have shown promise or achieved early success, and discuss the potential for scaling these techniques in practical quantum computing workflows.

### 7.5.1 Learning classical data distributions

A major line of research has focused on using QCBMs, QGANs, and hybrid QVAEs to learn structured classical datasets. These include:

- **Binary images** such as bars-and-stripes or small MNIST digits,
- **Correlated patterns** from spin models or statistical physics, and
- **Synthetic tabular data** with controllable complexity.

In these cases, quantum models are trained to reproduce a classical probability distribution $p_{\text{data}}(\mathbf{x})$ by minimizing a divergence (e.g. KL or MMD). Because QCBMs encode the distribution directly into quantum amplitudes, they offer a natural advantage in representing highly entangled or non-factorizable correlations that may be difficult for shallow classical models.

For example, in a recent study [34], QCBMs were used to learn event distributions from simulated high-energy physics experimental data, demonstrating that quantum circuits with modest depth and hardware-efficient layout can successfully model nontrivial correlations present in collider observables.

### 7.5.2 Quantum state learning and tomography

QGMs can be used to learn and reconstruct unknown quantum states, an application that directly leverages their quantum-native architecture. The goal is to train a parameterized quantum circuit $U(\theta)$ such that its output state matches a target quantum state $\rho_{\text{target}}$, either in terms of measurement statistics or through a fidelity-based loss:

$$\mathbb{L}_{\text{fidelity}} = 1 - F(\rho_{\theta}, \rho_{\text{target}}), \quad \text{where} \quad F(\rho, \sigma) = \left(\text{Tr}\sqrt{\sqrt{\rho}\ \sigma\ \sqrt{\rho}}\right)^2. \tag{7.17}$$

This approach by-passes full state tomography and has been shown to scale better for higher-dimensional systems [35]. It is particularly well-suited for applications in quantum chemistry, variational state preparation, or certifying outputs of quantum simulators

### 7.5.3 Quantum circuit compilation and data compression

The ability of QGMs to approximate arbitrary distributions or quantum states also enables applications in quantum data compression. For instance, once a target quantum state is learned via a QVAE, one can extract a latent representation that is lower-dimensional or more efficiently preparable. In this context:

- The **encoder** maps a high-complexity state to a latent space and
- The **decoder** reconstructs the state from fewer quantum resources.

This can be used to compress output states of variational quantum algorithms [36] or to recompile deep circuits into shallower ones with similar performance [37].

### 7.5.4 Anomaly detection

An emerging application of QGMs is in **quantum anomaly detection**, where the model is trained to learn a baseline distribution (either quantum or classical), and deviations from it are used to flag rare or unexpected inputs. These include:

- Quantum noise or hardware errors,
- Novel events in experimental data (e.g. in particle physics), and
- Out-of-distribution detection in classical data streams.

Fidelity-based metrics, expectation values of observables, or distance in latent space can all serve as **anomaly scores**. For example, QGANs have been tested for anomaly detection in high-energy physics, where quantum resources may help distinguish subtle features with fewer samples [38].

### 7.5.5 Physical simulation and sampling

QGMs have also been proposed as variational samplers in quantum Monte Carlo methods or for preparing input distributions in Hamiltonian simulation. Applications include:

- Sampling from Gibbs or Boltzmann distributions,
- Preparing ground states of physical Hamiltonians, and
- Modeling time-evolved states via QCBMs or quantum flows.

Recent work [39] has shown that **quantum phase estimation (QPE)** can be reinterpreted as a quantum generative model for sampling **dynamical response functions** of many-body systems. By encoding spectral densities into a quantum circuit, one can efficiently sample from distributions representing observables such as the dynamic structure factor, optical conductivity, and nuclear magnetic resonance (NMR) spectra. Table 7.3 presents a comparative overview of key application areas for quantum generative models. The table categorizes these use cases by their domain and identifies the most suitable QGM architectures for each

**Table 7.3.** Representative applications of quantum generative models across classical and quantum domains. Models include QCBMs, QGANs, QVAEs, and quantum flows.

| Application area | Model type | Task | Quantum advantage potential |
|---|---|---|---|
| Classical data modeling | QCBM, QGAN, QVAE | Reproduce $p_{\text{data}}(\mathbf{x})$ | Moderate (entanglement, efficiency) |
| Quantum state learning | QCBM, QGAN | Match $\rho_{\text{target}}$ | High (quantum-native, scalable) |
| Quantum circuit compression | QVAE | Reduce circuit depth/latent encoding | High (resource-efficient) |
| Anomaly detection | QCBM, QGAN | Detect deviations from learned prior | Moderate (sample efficiency) |
| Quantum simulation and sampling | QCBM, QBM | Variational sampling/distribution emulation | High (for complex physical models) |

task. It also provides qualitative assessment of the quantum advantage potential, based on the current literature and scaling behavior.

## 7.6 Open challenges and future directions

Quantum generative modeling presents both exciting prospects and substantial technical hurdles. As a rapidly evolving subfield of quantum machine learning, it sits at the intersection of quantum computing, probabilistic modeling, and optimization. This section discusses key **open challenges** currently limiting the scalability and practicality of quantum generative models, followed by a summary of **opportunities** for advancing the field, both theoretically and experimentally.

### 7.6.1 Challenges

Training quantum generative models often requires optimizing non-convex, high-dimensional landscapes. In variational settings, in particular with deep or unstructured ansatzes, gradients can vanish exponentially with the number of qubits. This severely limits scalability for circuits initialized randomly or without prior knowledge. Understanding the relationship between circuit architecture, expressivity, and learnability remains an open research area, particularly when targeting classically hard distributions.

QGMs are not immune to the failure modes that classical generative models encounter. For example, QGANs and QCBMs can suffer from **mode collapse**, where the model learns to generate only a subset of the target distribution's support. This is exacerbated by limited expressivity in shallow circuits or poor entangling strategies, and by the inability of some training objectives to fully penalize degeneracy.

Furthermore, gradient estimation on quantum hardware relies on parameter-shift rules or finite difference approximations, both of which require a large number of circuit evaluations and are sensitive to noise and sampling error.

QGMs, like many PQC-based models, are susceptible to **hardware noise**, including gate infidelity, readout error, and decoherence. Even as hardware fidelities improve, the reliance on finite samples leads to instabilities in training statistics or biased estimators in low-sample regimes. Shot noise affects loss estimation, gradient direction, and model convergence. Overall, these effects distort the output distribution and make training inefficient, in particular when high precision is required for tasks such as quantum state learning or anomaly detection.

### 7.6.2 Evaluation metrics and benchmarks

Unlike classical models, quantum distributions are not always easy to interpret or evaluate. Standard metrics such as KL divergence, inception score, or Fréchet inception distance may not be applicable or tractable. Fidelity, trace distance, and observable-based metric are useful, but often require access to the full target distribution or ideal output state.

In addition, certifying generative model performance in the quantum setting remains challenging, in particular when neither the data nor the model can be simulated classically. This limits cross-comparison between models and slows down the development of domain-specific techniques, such as for quantum chemistry, condensed matter, or high-energy physics.

While several proof-of-concept studies have demonstrated QGM training on toy problems, there is a lack of **standardized benchmarks** and **real-world datasets** tailored to quantum learning tasks. QGMs are well-positioned to address niche domains where classical generative methods struggle, such as sparse, structured data from nuclear physics, lattice gauge theories, sensor networks, or quantum tomography. However, in such contexts, the balance between noise and the quantum inductive bias must be carefully controlled—it may lead to more data-efficient or physically faithful models, or the lack of precision may inhibit training altogether.

### 7.6.3 Opportunities

Despite these challenges, quantum generative modeling offers a range of promising directions for theoretical advancement and near-term experimentation.

There is active work in designing expressive, trainable, and hardware-efficient ansatzes, including quantum convolutional and autoregressive circuits, quantum normalizing flows, and multi-scale entanglement structures. These architectures aim to balance depth, trainability, and expressivity while being compatible with real devices.

Several tasks in quantum physics—such as generating samples from spectral functions, learning many-body wavefunctions, or sampling from high-rank observables—may offer provable or practical quantum advantage. These settings are particularly amenable to QGMs since the sampling task itself is quantum-native and intractable classically.

New training paradigms, including hybrid quantum–classical workflows and non-adversarial loss functions (e.g. fidelity-based, score-matching, or kernel-based), offer more stable convergence than adversarial schemes. They also make better use of classical computing for parameter updates and gradient smoothing. Additionally, incorporating noise models or injecting noise during training loop may produce quantum models that are robust to near-term hardware noise. These approaches need to incorporate noise-mitigation methods such as zero-noise extrapolation, probabilistic error cancellation, and robust optimization help stabilize training under realistic conditions. Again, this raises the challenge of building efficient, noise robust workflows.

## 7.7 Summary

QGMs offer a compelling new approach to probabilistic modeling, leveraging the unique capabilities of quantum mechanics to represent and learn complex distributions. From QCBMs to QGANs and hybrid variational autoencoders, a diverse landscape of architectures has emerged, each with distinct strengths, training methodologies, and potential applications.

In this chapter, we began by tracing the classical roots of generative modeling, particularly through Boltzmann machines and general energy-based models and showed how these ideas naturally extend into the quantum domain. We discussed how quantum circuits can encode probability distributions via the Born rule and explored different classes of models tailored for tasks ranging from classical data generation to quantum state learning and spectral sampling.

A key theme throughout the chapter was **expressivity:** the ability of quantum circuits to represent complex distributions more efficiently than their classical counterparts. We reviewed formal characterizations of expressivity and linked them to circuit depth, entanglement, and architectural design. We also examined practical considerations in **training** these models, including loss functions, gradient estimation, and the role of shot noise and hardware constraints.

Through various **applications and case studies**, we highlighted how QGMs can be used in domains such as anomaly detection, circuit compression, and physics-informed sampling. These examples illustrated both the promise and the near-term feasibility of QGMs when deployed on current or upcoming quantum hardware.

At the same time, we addressed persistent **challenges** that remain critical barriers to scale. Yet the field is rapidly advancing, and emerging techniques in noise-aware training, hybrid optimization, and expressive circuit design continue to push the boundaries of what is achievable.

Looking forward, we anticipate that quantum generative modeling will play a foundational role in quantum machine learning and simulation. Ultimately, QGMS exemplify the hybrid paradigm of quantum learning: uniting the statistical power of generative modeling with the expressive and computational resources of quantum circuits. Their development will be pivotal to realizing practical quantum advantage in machine learning.

## References

[1] Sherrington D and Kirkpatrick S 1975 Solvable model of a spin-glass *Phys. Rev. Lett.* **35** 1792–6

[2] Hinton G E and Sejnowski T J 1983 Optimal perceptual inference *Proc. IEEE Conf. on Computer Vision and Pattern Recognition* vol 448 *(Washington, DC)* pp 448–53

[3] Ackley D H, Hinton G E and Sejnowski T J 1985 A learning algorithm for Boltzmann machines *Cogn. Sci.* **9** 147–69

[4] Koller D and Friedman N 2009 *Probabilistic Graphical Models: Principles and Techniques* (Cambridge, MA: MIT Press)

[5] Dempster A P, Laird N M and Rubin D B 1977 Maximum likelihood from incomplete data via the *EM* algorithm *J. R. Stat. Soc.* B **39** 1–38

[6] Pearl J 2022 Reverend Bayes on inference engines: a distributed hierarchical approach *Probabilistic and Causal Inference: The Works of Judea Pearl* (New York: Association for Computing Machinery) pp 129–38

[7] Hinton G E 2012 A practical guide to training restricted Boltzmann machines *Neural Networks: Tricks of the Trade* 2nd edn (Berlin: Springer) pp 599–619

[8] Hinton G E, Osindero S and Teh Y-W 2006 A fast learning algorithm for deep belief nets *Neural Comput.* **18** 1527–54

[9] Hinton G E 2002 Training products of experts by minimizing contrastive divergence *Neural Comput.* **14** 1771–800

[10] Teh Y W, Welling M, Osindero S and Hinton G E 2003 Energy-based models for sparse overcomplete representations *J. Mach. Learn. Res.* **4** 1235–60

[11] Goodfellow I, Bengio Y and Courville A 2016 *Deep Learning* (Cambridge, MA: MIT Press)

[12] Goodfellow I J, Pouget-Abadie J, Mirza M, Xu B, Warde-Farley D, Ozair S, Courville A and Bengio Y 2014 Generative adversarial networks arXiv: 1406.2661

[13] Papamakarios G, Nalisnick E, Rezende D J, Mohamed S and Lakshminarayanan B 2021 Normalizing flows for probabilistic modeling and inference *J. Mach. Learn. Res.* **22** 2617–80

[14] Das A and Chakrabarti B K 2008 Colloquium: quantum annealing and analog quantum computation *Rev. Modern Phys.* **80** 1061–81

[15] Amin M H, Andriyash E, Rolfe J, Kulchytskyy B and Melko R 2018 Quantum Boltzmann machine *Phys. Rev.* X **8** 021050

[16] King J, Yarkoni S, Raymond J, Ozfidan I, King A D, Nevisi M M, Hilton J P and McGeoch C C 2019 Quantum annealing amid local ruggedness and global frustration *J. Phys. Soc. Jpn.* **88** 061007

[17] Benedetti M, Garcia-Pintos D, Perdomo O, Leyton-Ortega V, Nam Y and Perdomo-Ortiz A 2019 A generative modeling approach for benchmarking and training shallow quantum circuits *npj Quantum Inf* **5** 45

[18] Liu J-G and Wang L 2018 Differentiable learning of quantum circuit Born machines *Phys. Rev.* A **98** 062324

[19] Coyle B, Mills D, Danos V and Kashefi E 2020 The Born supremacy: quantum advantage and training of an ising Born machine *npj Quantum Inf.* **6** 60

[20] Gretton A, Borgwardt K M, Rasch M J, Schölkopf B and Smola A 2012 A kernel two-sample test *J. Mach. Learn. Res.* **13** 723–73

[21] Mohamed S and Lakshminarayanan B 2016 Learning in implicit generative models arXiv: 1610.03483

[22] Dallaire-Demers P-L and Killoran N 2018 Quantum generative adversarial networks *Phys. Rev.* A **98** 012324

[23] Romero J, Olson J P and Aspuru-Guzik A 2017 Quantum autoencoders for efficient compression of quantum data *Quantum Sci. Technol.* **2** 045001

[24] Larocca M, Thanasilp S, Wang S, Sharma K, Biamonte J, Coles P J, Cincio L, McClean J R, Holmes Z and Cerezo M 2025 Barren plateaus in variational quantum computing *Nat. Rev. Phys.* **7** 174–89

[25] Pirnay N, Sweke R, Eisert J and Seifert J-P 2023 Superpolynomial quantum-classical separation for density modeling *Phys. Rev.* A **107** 042416

[26] Sim S, Johnson P D and Aspuru-Guzik A 2019 Expressibility and entangling capability of parameterized quantum circuits for hybrid quantum-classical algorithms *Adv. Quantum Technol.* **2** 1900070

[27] Du Y, Hsieh M-H, Liu T and Tao D 2020 Expressive power of parametrized quantum circuits *Phys. Rev. Res.* **2** 033125

[28] Renes J M, Blume-Kohout R, Scott A J and Caves C M 2004 Symmetric informationally complete quantum measurements *J. Math. Phys.* **45** 2171–80

[29] Gross D, Audenaert K and Eisert J 2007 Evenly distributed unitaries: on the structure of unitary designs *J. Math. Phys.* **48** 052104

[30] Delgado A, Rios F and Hamilton K E 2023 Identifying overparameterization in quantum circuit Born machines arXiv: 2307.03292

[31] Zoufal C 2021 Generative quantum machine learning arXiv: 2111.12738

[32] Csiszar I 1975 $i$ -divergence geometry of probability distributions and minimization problems *Ann. Probab.* **3** 146–58

[33] Gretton A, Borgwardt K, Rasch M J, Scholkopf B and Smola A J 2008 A kernel method for the two-sample problem arXiv: 0805.2368

[34] Delgado A and Hamilton K E 2022 Unsupervised quantum circuit learning in high energy physics *Phys. Rev.* D **106** 096006

[35] Innan N, Siddiqui O I, Arora S, Ghosh T, Koçak Y P, Paragas D, Al Omar Galib A, Al-Zafar Khan M and Bennai M 2024 Quantum state tomography using quantum machine learning *Quant. Mach. Intell.* **6** 28

[36] Huang C-J, Ma H, Yin Q, Tang J-F, Dong D, Chen C, Xiang G-Y, Li C-F and Guo G-C 2020 Realization of a quantum autoencoder for lossless compression of quantum data *Phys. Rev.* A **102** 032412

[37] Quetschlich N, Kiwit F J, Wolf M A, Riofrio C A, Burgholzer L, Luckow A and Wille R 2024 Towards application-aware quantum circuit compilation arXiv: 2404.12433

[38] Bermot E, Zoufal C, Grossi M, Schuhmacher J, Tacchino F, Vallecorsa S and Tavernelli I 2023 Quantum generative adversarial networks for anomaly detection in high energy physics *2023 IEEE Int. Conf. Quantum Comput. Eng. (QCE)* (Piscataway, NJ: IEEE) pp 331–41

[39] Sels D and Demler E 2021 Quantum generative model for sampling many-body spectral functions *Phys. Rev.* B **103** 014301

**IOP** Publishing

Quantum Machine Learning
Concepts and possibilities
**Andrea Delgado and Kathleen E Hamilton**

# Chapter 8

# Theory, expressivity, and learning bounds

Machine learning theory gives us a language to reason about what a model can represent (expressivity), how many samples it needs to learn (sample complexity), how well it will perform on unseen data (generalization), and what the *fundamental limitations* are. Quantum machine learning inherits these questions and complicates them: hypothesis classes now live in exponentially large Hilbert spaces, training dynamics may be governed by quantum Fisher information metrics rather than Euclidean ones, and access models to the data (classical versus quantum states, labeled versus unlabeled, membership queries versus i.i.d. samples) crucially affect what can be learned efficiently. This chapter develops a self-contained framework to address these issues:

1. **Expressivity.** We formalize quantum hypothesis classes (e.g. variational quantum circuits, quantum kernels, quantum generative models) and compare them to classical counterparts using measures such as pseudo-dimension, covering numbers, Rademacher complexity, effective dimension/kernel eigenvalue decay, entanglement capacity, circuit depth/2-designness, and the rank of induced feature maps.
2. **Learning bounds.** We adapt the probably approximately correct (PAC) and agnostic learning notions to quantum settings (both with classical and quantum data) and collect known upper and lower bounds on sample complexity and generalization error. We stress when quantum examples (quantum superpositions of labeled samples) do or do not change the asymptotics, and distinguish between information-theoretic and computational advantages.
3. **Quantum learning theory results.** We summarize results showing (i) no super-polynomial reductions in PAC sample complexity from quantum examples (beyond constant/low-order factors), (ii) provable quantum speedups in query or time complexity for specific concept classes or under certain data-access models, and (iii) constructions of quantum kernels or feature maps

doi:10.1088/978-0-7503-4952-9ch8  

that can be exponentially hard to emulate classically (subject to complexity assumptions), along with their generalization analyses.

4. **Limitations and no-go theorems.** We discuss hardness-of-training results, dequantization of several linear-algebraic quantum machine learning algorithms, barren plateaus, expressivity–trainability trade-offs, stability/robustness limits under noise, and finite measurement budgets.
5. **Open problems.** We close with a research agenda: tighter, noise-aware generalization bounds for variational circuits; a unifying complexity-theoretic taxonomy of quantum hypothesis classes; PAC-Bayes, compression, and algorithmic stability frameworks for quantum machine learning; and principled ways to relate entanglement structure, kernel spectra, and trainability.

We assume familiarity with basic quantum computing (states, unitaries, measurements—chapter 2), introductory statistical learning theory (PAC learning, Vapnik–Chervonekis (VC) dimension, generalization bounds), and elementary functional analysis. The chapter is otherwise self-contained and provides boxed definitions and theorems for quick reference.

## 8.1 Definitions and frameworks for expressivity

This section establishes rigorous foundations for comparing the expressivity of classical and quantum hypothesis classes. By articulating definitions and measures used to quantify model complexity and representational power, we lay the groundwork for understanding the theoretical performance and limitations of quantum machine learning models.

### 8.1.1 Hypothesis classes

In statistical learning theory, a hypothesis class $\mathbf{H}$ is a set of functions mapping inputs to outputs. In classical machine learning, common hypothesis classes include neural networks, decision trees, linear separators, and kernel-based models. These models operate in classical computational paradigms, and their expressive capacity is typically characterized by parameters such as the number of layers and neurons (in neural networks), the tree depth (in decision trees), or the norm bounds and kernel choices (in kernel methods).

In contrast, quantum hypothesis classes are defined by quantum circuits that act on quantum states and are measured according to some observable. The quantum analog of a hypothesis class is typically constructed using a parameterized quantum circuit (PQC) or variational quantum circuit that encodes classical data into quantum states, processes them through a unitary transformation, and measures them to produce an output prediction.

**Quantum hypothesis class**
Let $\mathscr{H}_Q$ be a class of quantum hypotheses defined as

$$\mathscr{H}_Q = \{h_\theta(x) = \mathbb{E}[M \mid U_\theta U_{\text{enc}}(x)|0\rangle]: \theta \in \Theta\},$$

where $U_{\text{enc}}(x)$ is a unitary encoding of the classical input $x$, $U_\theta$ is a variational quantum circuit parametrized by $\theta$, and $M$ is a Hermitian operator corresponding to a projective or positive operator-valued measure (POVM) measurement. The output $h_\theta(x)$ is the expected value of the observable $M$ with respect to the quantum state $U_\theta U_{\text{enc}}(x)|0\rangle$.

This formulation captures a wide variety of quantum machine learning models, including variational classifiers (chapter 6), quantum generative models (chapter 7), and quantum kernel methods (chapter 4). The precise structure of $U_\theta$ and the choice of measurement operator $M$ determines the expressivity and learnability of the hypothesis class.

### 8.1.2 Quantifying expressivity

To compare classical and quantum models in a principled manner, we require quantitative measures of expressivity. These measures often reflect the richness or capacity of the hypothesis class $\mathscr{H}$ to approximate a wide variety of functions. Several frameworks exist to capture this notion, including complexity-theoretic, geometric, and spectral metrics.

#### *8.1.2.1 Classical complexity measures*

In classical learning theory, the following notions are widely used:

- The **Vapnik–Chervonekis (VC) dimension** [1], which measures the largest number of points that can be shattered (correctly labeled in all possible ways) by the hypothesis class.
- The **pseudo-dimension** [2], an extension of the VC dimension to real-valued functions.
- the **Rademacher complexity**, which quantifies the ability of a hypothesis class to fit random noise.
- **Covering numbers** and **fat-shattering dimension**, which describe the metric entropy and granularity of function classes.
- **Compression schemes** and **algorithmic stability**, which are relevant for bounding generalization error.

#### *8.1.2.2 Kernel-based metrics*

For kernel machines, including quantum kernel methods, a widely used metric is the **effective dimension**.

**Effective dimension**

Let $K \in \mathbb{R}^{n \times n}$ be the kernel matrix associated with a hypothesis class, and $\lambda > 0$ a regularization parameter. The effective dimension is defined as

$$d_{\text{eff}}(\lambda) = \text{Tr}(K(K + \lambda I)^{-1}).$$

This quantity serves as a capacity measure that often replaces the VC dimension in kernelized learning bounds. It depends on the spectral decay of $K$ and characterizes the sample complexity of the model.

For quantum kernels, $K$ may be constructed using a quantum feature map $\phi(x) = U_\phi(x)|0\rangle$, leading to entries $K_{ij} = |\langle \phi(x_i)\phi(x_j)\rangle|^2$. When the eigenvalues of $K$ decay rapidly, the effective dimension remains small, enabling generalization despite the high-dimensional Hilbert space.

*8.1.2.3 Quantum-specific expressivity measures*

Quantum models admit additional metrics that are absent in classical settings:

- **Entangling power** measures the ability of a variational circuit to generate entanglement. Highly entangled states typically expand the expressivity of the model but may be harder to optimize and generalize from.
- **Proximity to a unitary t-design**, which assesses how well a circuit approximates a random unitary ensemble. Circuits that approach a 2-design, for instance, exhibit concentration of measure phenomena.
- The **quantum Fisher information matrix** (QFIM), which governs the local sensitivity of quantum states to parameter changes. Its rank and conditioning can be used to estimate the dimensionality of the effective model manifold.
- **Circuit depth and locality**, which constrain the types of correlations that a model can express. Shallow circuits with local connectivity have limited expressivity.

Figure 8.1 illustrates the conceptual landscape of classical and quantum hypothesis classes. The ellipses represent different families of models used in machine learning, including classical kernel methods, quantum kernels, variational quantum circuits (PQCs or quantum neural networks (QNNs)), and quantum generative models such as Born machines. The overlapping regions reflect shared representational capabilities, while the distinct regions indicate expressivity that may be inaccessible to other model types.

The labeled intersections (e.g. 'Born Machines') highlight areas where the expressive power of quantum models is conjectured—although not always rigorously proven—to exceed that of classical models under reasonable complexity-theoretic assumptions.

### 8.1.3 Expressivity versus trainability

While high expressivity is generally desirable for representing complex functions, it can negatively impact trainability. This tension is particularly pronounced in

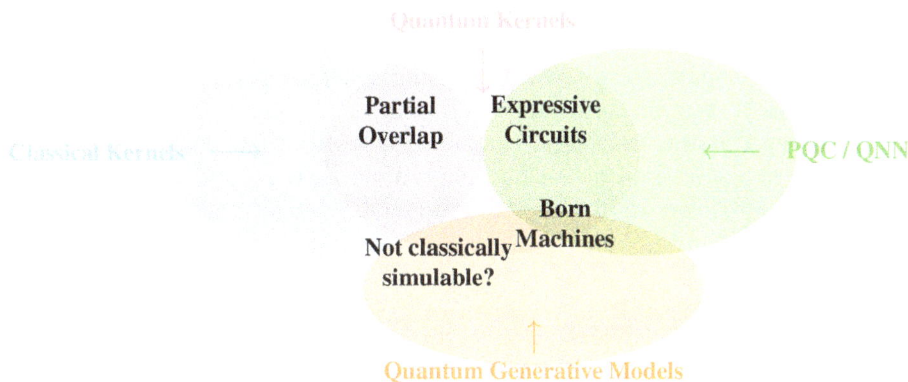


**Figure 8.1.** Conceptual map of classical and quantum hypothesis classes. Quantum kernel methods and variational quantum circuits (PQC/QNN) can represent functions outside classical reach. Quantum generative models such as Born machines may further push the boundary of expressivity.

quantum models, where the curse of dimensionality and vanishing gradients can hinder optimization.

One manifestation of this issue is the **barren plateau phenomenon**, where the variance of the gradient of a loss function decays exponentially with the number of qubits. This makes it extremely difficult to train deep or highly expressive variational circuits, as small gradients are quickly overwhelmed by noise or finite sampling error.

**Barren plateau theorem (sketch)**
Let $U(\theta)$ be a variational circuit sampled from a distribution approximating a unitary 2-design. Then, for a randomly initialized circuit of depth $d$ acting on $n$ qubits, the variance of the gradient of a loss function $L(\theta)$ with respect to $\theta$ satisfies

$$\mathrm{Var}(\nabla_\theta L) = \mathcal{O}\left(\frac{1}{\mathrm{poly}(2^n)}\right).$$

This result implies that the gradients vanish exponentially in the system size, making gradient-based optimization impractical without additional structure.

Several mitigation strategies have been proposed, including layerwise training, local cost functions, and restricted ansatz families that avoid expressivity levels leading to such plateaus. Nevertheless, designing quantum models that strike the right balance between expressivity and trainability remains a central challenge in the field.

In the following section, we build on this framework to examine how expressivity relates to the learnability of quantum models, quantifying the sample complexity and generalization properties associated with each class.

## 8.2 Learning performance: sample complexity and generalization

This section investigates how expressive quantum hypothesis classes translate into learnable models. We develop bounds on sample complexity and generalization error, adapting classical learning-theoretic concepts to quantum settings. We also highlight the subtleties introduced by quantum-data access and quantum circuits, and summarize known results on both upper and lower bounds for quantum learning.

### 8.2.1 PAC and agnostic learning

In the probably approximately correct (PAC) learning model [3, 4], the learner aims to find a hypothesis with lower generalization error using only a limited number of training samples drawn i.i.d. from a fixed distribution. Let $\mathscr{H}$ be a hypothesis class mapping inputs $x \in \mathscr{X}$ to labels $y \in \{0, 1\}$ or real values in regression. The learner has access to a dataset $S = \{(x_i, y_i)\}_{i=1}^{m} \sim \mathscr{D}^m$, and the goal is to find $h \in \mathscr{H}$ such that the population risk $R(h)$ is close to the minimal possible risk within $\mathscr{H}$:

$$R(h) = \mathbb{E}_{(x,\, y)\sim\mathscr{D}}[l(h(x),\, y)], \tag{8.1}$$

where $l$ is a loss function such as $0 - 1$ loss, squared loss, or cross-entropy.

In the **agnostic** learning model, we do not assume the data labels are consistent with any hypothesis in $\mathscr{H}$. The learner instead competes with the best possible hypothesis in the class.

Quantum models can be analysed in this framework with minor adjustments. The hypothesis $h_\theta$ now results from a measurement on a quantum circuit:

$$h_\theta(x) = \mathbb{E}[M \,|\, U_\theta U_{\text{enc}}(x)|0\rangle\, ], \tag{8.2}$$

where the expectation is taken with respect to the Born rule over repeated shots. The training dataset may be classical (a list of $(x_i, y_i)$) or quantum, consisting of access to pure or mixed quantum states encoding the data–label pairs.

Two quantum-data models are often considered:

- **Classical access:** The learner receives $(x_i, y_i)$ drawn i.i.d. from $\mathscr{D}$.
- **Quantum examples:** The learner has access to the quantum superposition stat,:

$$|\psi_{\mathscr{D}}\rangle = \sum_{x,y} \sqrt{p(x,\, y)}\, |x,\, y\rangle, \tag{8.3}$$

possibly along with oracle access to related operations.

The PAC learning guarantees under both access models often follow similar forms, with dependence on a complexity measure such as VC dimension or effective dimension.

**PAC learning bound (classical)**
Let $\mathscr{H}$ be a hypothesis class of finite VC dimension $d$, and let $\delta \in (0, 1)$. Then, with probability at least $1 - \delta$ over $m$ training examples sampled i.i.d. from $\mathscr{D}$, every hypothesis $h \in \mathscr{H}$ satisfies

$$R(h) \leqslant \hat{R}_S(h) + \mathcal{O}\left(\sqrt{\frac{d\log(m/d) + \log(1/\delta)}{m}}\right),$$

where $\hat{R}_S(h)$ is the empirical risk over the sample $S$.

These bounds extend to quantum learners when hypothesis classes are suitably restricted (e.g. finite precision, bounded output) and the loss is estimated finite-shot measurements.

### 8.2.2 Quantum sample complexity

It is natural to ask whether quantum learning models offer improvements in sample complexity, that is, do they require fewer training examples to achieve the same generalization error? The answer depends critically on the data-access model.

In the **standard PAC model with classical examples**, it is known that quantum models do not enjoy asymptotically better sample complexity than classical models. That is, the number of examples needed to learn a class $\mathscr{H}$ up to error $\varepsilon$ is lower bounded by $\Omega(d/\varepsilon^2)$, where $d$ is the VC dimension or effective dimension of $\mathscr{H}$.

However, when learners have access to quantum examples they may experience constant-factor improvements in sample complexity or stronger query complexity reductions in specific oracle-based tasks. These do not violate known lower bounds in classical PAC learning but represent separations in models of computation.

**Quantum PAC learnability result**
Let $\mathscr{H}$ be a concept class learnable with sample complexity $m(\varepsilon, \delta)$ in the classical PAC model. Then, under access to quantum examples $|\psi_{\mathscr{D}}\rangle$, $\mathscr{H}$ remains PAC learnable with sample complexity

$$m_Q(\varepsilon, \delta) = \mathcal{O}(m(\varepsilon, \delta)),$$

with potential constant-factor improvements depending on the concept class. However, no super-polynomial separation in sample complexity is possible under standard assumptions.

Several concept classes exhibit provable separations in query complexity when learners are given quantum membership or evaluation oracles. For example:

- Learning Boolean juntas with fewer queries [5].
- Fourier-sparse functions [6].
- Parity functions with noise [7].

### 8.2.3 Generalization bounds for quantum models

Generalization bounds quantify how well a model trained on finite data will perform on unseen data. For quantum models, classical learning theory provides the blueprint, but it must be adapted to account for:

- The nature of quantum hypothesis spaces (e.g.. variational circuits, kernel-induced Hilbert spaces)
- Shot noise in finite sampling
- Non-Euclidean optimization geometry

*8.2.3.1 Quantum kernel models*

For quantum kernel methods, generalization error is governed by the effective dimension $d_{\text{eff}}(\lambda)$, which depends on the spectrum of the kernel matrix. Specifically, for regularized empirical risk minimization (e.g. kernel ridge regression), the following bound holds.

**Generalization bound via effective dimension**

Let $K$ be a quantum kernel matrix on $m$ training examples and $\lambda > 0$ the regularization parameter. Then the excess risk of the trained predictor satisfies

$$R(h) - R(h^*) \leqslant \mathcal{O}\left(\frac{d_{\text{eff}}(\lambda)}{m}\right) + \mathcal{O}(\lambda \, |f^*|^2),$$

where $f^*$ is the target function in the reproducing kernel Hilbert space (RKHS) and

$$d_{\text{eff}}(\lambda) = \text{Tr}(K(K + \lambda I)^{-1}).$$

This result reveals that high-dimensional quantum feature maps do not automatically result in overfitting, generalization depends instead on the **spectral decay** of the kernel.

*8.2.3.2 Variational quantum circuits*

For variational models, generalization can be bounded using:

- **Uniform stability** of the training algorithm.
- **PAC-Bayes bounds** on the classical post-measurement distribution.
- **Quantum Fisher information** geometry to assess sensitivity to parameters

Table 8.1 summarizes representative generalization bounds across various classical and quantum learning models. For classical and quantum kernel methods, generalization performance is controlled by the effective dimension $d_{\text{eff}}(\lambda)$, which captures the complexity of the kernel matrix via its spectrum. When the kernel eigenvalues decay rapidly, these models can generalize well even in high-dimensional Hilbert spaces.

**Table 8.1.** Representative generalization bounds for classical and quantum hypothesis classes. Quantum models inherit similar sample complexity behavior under well-behaved spectra or stable training dynamics, but some variational models suffer from barren plateaus that hinder generalization.

| Model type | Complexity measure | Generalization bound | Reference |
|---|---|---|---|
| Classical kernel (ridge) | Effective dim. $d_{\text{eff}}(\lambda)$ | $\mathcal{O}\left(\frac{d_{\text{eff}}(\lambda)}{m}\right)$ | Steinwart *et al* [8] |
| Quantum kernel | $d_{\text{eff}}(\lambda)$, spectral decay | $\mathcal{O}\left(\frac{d_{\text{eff}}(\lambda)}{m}\right)$ | Schuld *et* al [9] |
| PQC/QNN (stochastic training) | Training stability, QFIM | PAC-Bayes-type bound on measurement-induced distributions | Abbas *et al* [10] |
| PQC (global cost functions) | Gradient norm, barren plateau effect | No meaningful bound (exponential gradient variance) | McClean *et al* [11] |

Variational quantum models (e.g. PQCs or QNNs) are more challenging to analyse directly using classical complexity measures. Instead, generalization bounds are often obtained through algorithmic stability arguments or PAC-Bayesian frameworks applied to the output distributions of quantum measurements. Some bounds also rely on the structure of the QFIM, which governs the circuit's sensitivity to parameters.

As shown in the final row of the table, models with poorly conditioned cost landscapes lack meaningful generalization guarantees, since gradients vanish exponentially and training cannot reliably converge. This highlights the need for both expressive and trainable architectures when designing quantum learning models.

## 8.3 Positive results from quantum learning

This section highlights provable contexts where quantum learners offer advantages over classical counterparts. These advantages can manifest in reduced query complexity, improved computational efficiency, or the ability to realize hypothesis classes that are hard to emulate classically.

### 8.3.1 Query and time complexity speedups

Several results from quantum learning theory establish that, under specific oracle models or problem instances, quantum algorithms can outperform classical ones in terms of **query complexity** or **runtime**. These separations are typically shown in idealized models such as access to quantum membership or evaluation oracles.

#### *8.3.1.1 Learning juntas*

In the classical setting, learning a $k$–junta (a Boolean function depending on $k \ll n$ out of $n$ variables) requires $\Omega(k \log n)$ queries in the worst case. Quantum algorithms can reduce this to $\mathcal{O}(k)$ using Fourier sampling or Grover-type searches [5].

*8.3.1.2 Fourier-sparse functions*
Concept classes that are sparse in the Fourier domain (i.e. with only a few nonzero Fourier coefficients) are more accessible to quantum learners via quantum Fourier sampling. In particular, estimating significant Fourier coefficients can be accomplished with quadratically fewer samples.

*8.3.1.3 Learning parity with noise*
Quantum learners can learn parity functions in the presence of random classification noise exponentially faster than any known classical *statistical query* (SQ) algorithm, due to their ability to leverage quantum interference [7].

These separations demonstrate that quantum learners may access different information-theoretic pathways through their interaction with oracles, even when their PAC sample complexity does not improve asymptotically.

### 8.3.2 Quantum kernels and feature maps

Quantum kernels are a prominent approach in quantum machine learning where classical data $x \in \mathbb{R}^d$ are encoded into quantum states via a unitary feature map $U_\phi(x)$, and learning is performed via inner products of the resulting states. The kernel is defined as

$$k(x, x') = |\langle \phi(x)|\phi(x')|^2, \tag{8.4}$$

where $|\phi(x)\rangle = U_\phi(x)|0\rangle$.

In some cases, these quantum kernels can be hard to simulate or estimate classically. This hardness is typically based on assumptions from quantum computational complexity (e.g. the hardness of simulating certain circuits implies classical intractability).

One example is the **quantum kernel estimation (QKE)** approach, where the kernel entries are estimated on a quantum device and then used to train a classical model such as a support vector machine (SVM). If the quantum kernel corresponds to a circuit class that implements a task believed to be classically intractable, then the associated hypothesis class may offer computational advantages over classical kernels.

However, provable generalization still depends on the effective dimension or alignment between the quantum kernel and the target function. That is, expressivity alone is not sufficient, good generalization depends on how well the kernel aligns with the structure of the data.

### 8.3.3 Quantum generative models

Quantum generative models aim to generate classical or quantum data by sampling from quantum circuits. These models have several theoretical advantages:

**Expressivity:**

- Quantum generative models can represent probability distributions that require exponentially many parameters to represent classically.
- Some quantum circuit Born machines (QCBMs) are shown to be universal generators under finite precision assumptions [12].

**Sample complexity and compression:**

- In hybrid quantum–classical training schemes, generative models can compress structured data into low-parameter quantum circuits that still achieve good coverage over the target distribution.

**Provable hardness of classical simulation:**

- Sampling from the output of specific QCBMs (e.g. instantaneous quantum polynomial-time (IQP) or random circuits with certain depth and entanglement structure) is conjectured to be classically hard under plausible assumptions (e.g. anticoncentration and average-case hardness).

This combination of expressivity, compactness, and complexity-theoretic hardness supports the view that quantum generative models could outperform classical ones in appropriate tasks, such as simulating quantum distributions or compressing large datasets into variational circuits.

Figure 8.2 presents a conceptual flowchart outlining the distinct stages of quantum machine learning workflow where quantum advantage may arise. The figure begins with **quantum access to data**, which includes tasks such as preparing

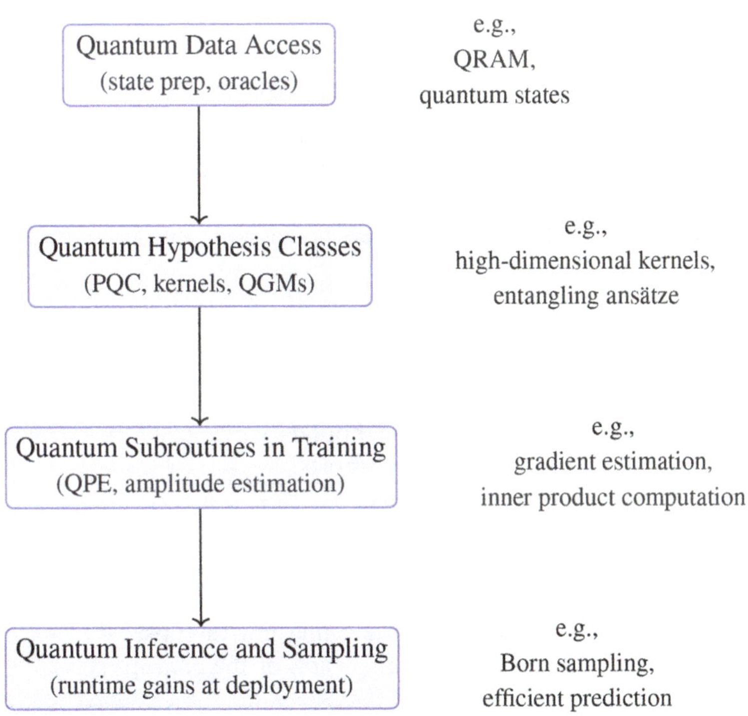


**Figure 8.2.** Schematic flowchart showing where quantum advantage may arise in a learning pipeline: from access to quantum data, through expressive model classes and quantum-accelerated training routines, to deployment-time inference or sampling advantages.

quantum states that encode classical inputs or accessing oracular quantum information. This is the foundational layer enabling quantum machine learning to operate in a regime beyond classical capabilities.

The second stage highlights the role of **quantum-enhanced hypothesis classes**, such as variational quantum circuits (PQCs), quantum kernel models, and quantum generative models (QGMs). These classes may have superior expressive power relative to classical models, potentially allowing them to capture patterns or distributions that are hard to model classically.

The third layer identifies opportunities for **quantum subroutines in training**, where algorithms such as quantum phase estimation (QPE) or amplitude estimation can accelerate optimization, gradient evaluation, or inner product computation. These subroutines could yield computational savings during model fitting, in particular when evaluating global properties of the model or data.

Finally, the bottom layer of the flowchart represents **quantum inference or sampling**, where quantum circuits are used at deployment to generate predictions or samples more efficiently than any known classical method. For instance, Born machines may generate samples from distributions that are provably hard to simulate classically, offering a clear advantage in generative modeling tasks.

Together, these four layers illustrate that quantum advantage in machine learning is not confined to a single aspect but can span data access, model class, training dynamics, and inference, depending on the algorithm and application.

## 8.4 Limitations, no-go theorems, and dequantization

While quantum machine learning holds significant promise, there are important theoretical and practical limits. This section provides a balanced view by discussing known lower bounds, structural limitations of variational circuits, dequantization results, and the fragility of quantum models under noise and finite measurements.

### 8.4.1 Sample complexity lower bounds

In classical learning theory, lower bounds on sample complexity are typically expressed in terms of VC dimension, Rademacher complexity, or covering numbers. In the quantum setting, many of these bounds carry over, in particular under classical data access or even quantum examples.

In particular, it has been shown [3] that:

- For general concept classes, access to quantum examples does not asymptotically reduce PAC sample complexity beyond logarithmic or constant-factor improvements.
- Lower bounds on quantum PAC learning mirror those of classical learners when measured against the VC dimension of the concept class.

Thus, while quantum access may help with query efficiency or runtime, it does not circumvent the fundamental need for a sufficient number of samples.

### 8.4.2 Barren plateaus and optimization barriers

A significant limitation of variational quantum algorithms is the occurrence of barren plateaus, regions in parameter space where the gradient of the cost function vanishes exponentially with the number of qubits.

There are several known types of barren plateaus:

- **Depth-induced barren plateaus:** Arise in deep unstructured ansatzes that from approximate 2-designs [11].
- **Noise-induced barren plateaus:** Caused by decoherence and finite sampling effects, even in shallow circuits [13].
- **Problem-induced barren plateaus:** Occur when the cost function has an unfavorable landscape, even for structured ansatzes [14].

**Gradient variance decay in PQCs**

Let $U(\theta)$ be a parametrized quantum circuit of depth $d$ acting on $n$ qubits. If $U(\theta)$ forms an approximate 2-design and $M$ is a local observable, then

$$\mathrm{Var}(\nabla_\theta \mathbb{E}[M]) = \mathcal{O}\left(\frac{1}{\mathrm{poly}(2^n)}\right),$$

implying that gradients vanish exponentially as the number of qubits increases.

These plateaus make optimization challenging and place practical limits on the scalability of variational quantum circuits without architecture-aware design or gradient-free methods.

### 8.4.3 Dequantization results

Several algorithms initially proposed as quantum enhancements have later been shown to admit efficient classical analogs under slightly modified assumptions. This process is referred to as **dequantization**. Examples include:

- **HHL algorithm variants:** Quantum linear system solvers inspired early quantum kernel methods, but under data-access assumptions (e.g. row-query or sampling access), classical solvers can achieve similar performance.
- **Quantum recommendation systems:** Tang [15] provided a classical algorithm with similar performance to a quantum recommendation algorithm based on linear algebra primitives.
- **Kernel methods:** Quantum kernels using feature maps with classically simulable circuits can often be emulated by classical methods without significant overhead.

These dequantization results emphasize the importance of specifying the data access model and complexity assumptions when claiming quantum speedup

### 8.4.4 Robustness and noise sensitivity

In practical quantum machine learning implementations, noise and limited measurements pose significant limitations:

- **Shot noise:** Finite sampling leads to uncertainty in gradient and loss estimates, in particular when using deep circuits or many parameters.
- **SPAM errors:** State preparation and measurement errors can bias outcomes and destabilize training.
- **Expressivity versus stability:** Highly expressive quantum models may be more sensitive to noise, exhibiting worse generalization when model complexity is not well-regularized.

## 8.5 Open problems and future directions

The theoretical foundations of quantum machine learning are rapidly evolving, and many important questions remain open. This section highlights promising directions for future research that aim to deepen our understanding of expressivity, generalization, and optimization in quantum models.

One of the key frontiers is the development of noise-aware generalization theory. While existing bounds often assume idealized conditions such as noiseless measurements and perfect state preparation, real quantum hardware introduces various sources of imperfection. Finite-shot sampling, gate errors, and state preparation and measurement (SPAM) noise affect both the accuracy of learning and the validity of theoretical guarantees. Addressing this gap will require generalization bounds that explicitly account for measurement variance, incorporate models of quantum noise, and extend stability-based and PAC-Bayes methods to realistic device settings.

A second major challenge is the lack of a unified taxonomy for quantum hypothesis classes. Unlike classical learning theory, which benefits from well-established complexity measures such as VC dimension and covering numbers, quantum models span a wide range of circuit classes and feature maps, many of which are not well understood in terms of their classical simulability or learning complexity. Identifying precise relationships among quantum kernels, PQCs, and classical analogs is essential for characterizing what quantum models can learn efficiently and how they differ in expressive power.

Another compelling direction is the study of optimization geometry. Variational quantum circuits exhibit landscapes that are governed by non-Euclidean geometry, particularly the structure induced by the QFIM. Exploiting this geometry may lead to better optimization algorithms, such as quantum natural gradient methods, and may offer insight into the generalization properties of variational models. Understanding the role of flat minima, QFI spectra, and parameter-space curvature could reveal new trainability–generalization trade-offs.

Related to this, there is the question of whether quantum models can admit PAC-Bayes-type generalization bounds. Classical models benefit from strong performance guarantees based on compression or probabilistic priors over hypothesis classes. Extending these techniques to quantum circuits requires defining meaningful priors over parameterized unitaries and bounding divergences between quantum measurement distributions. Such work could also guide the development of hybrid quantum–classical learners with robust generalization properties.

A critical challenge facing expressive quantum models is the onset of barren plateaus. This raises fundamental questions about the tradeoff between expressivity and trainability. Highly expressive models may have sufficient capacity to learn complex distributions, yet may be untrainable without architectural constraints. Research is needed to identify the conditions under which expressive models remain trainable and to design ansatz families that balance circuit depth, entanglement, and gradient flow.

Beyond modeling classical data, quantum learning theory must also address the task of learning from quantum data, states, channels, and measurements that arise in quantum physics and information processing. Theoretical tools such as shadow tomography and randomized measurement protocols offer pathways to learning in this setting, but a more rigorous understanding of the sample complexity, generalization, and trainability of quantum-data learners is required.

Finally, most existing theory focuses on worst-case guarantees. However, practical quantum machine learning applications are often structured, and real-world data exhibit regularity. This opens opportunities for distribution-dependent analysis, smoothed complexity bounds, and average-case generalization guarantees. Developing such theory could bridge the gap between quantum advantage in principle and in practice.

## References


[1] Vapnik V N and Chervonenkis A Y 1971 On the uniform convergence of relative frequencies of events to their probabilities *Theory Probab. – Its Appl.* **16** 264–80

[2] Pollard D 1984 *Convergence of Stochastic Processes* (New York: Springer)

[3] Arunachalam S and de Wolf R 2017 Guest column: A survey of quantum learning theory *ACM SIGACT News* **48** 41–67

[4] Servedio R A and Gortler S J 2004 Equivalences and separations between quantum and classical learnability *SIAM J. Comput.* **33** 1067–92

[5] Atıcı A and Servedio R A 2007 Quantum algorithms for learning and testing juntas *Quantum Inf. Process.* **6** 323–48

[6] Bernstein E and Vazirani U 1997 Quantum complexity theory *SIAM J. Comput.* **26** 1411–73

[7] Cross A W, Smith G and Smolin J A 2015 Quantum learning robust against noise *Phys. Rev.* A **92** 012327

[8] Steinwart I and Christmann A 2008 *Support Vector Machines* (New York: Springer)

[9] Schuld M and Killoran N 2019 Quantum machine learning in feature Hilbert spaces *Phys. Rev. Lett.* **122** 040504

[10] Abbas A and Andreopoulos Y 2022 PAC-Bayesian bounds on rate-efficient classifiers *Proc. Machine Learning Research* **162** 1–9

[11] McClean J R, Boixo S, Smelyanskiy V N, Babbush R and Neven H 2018 Barren plateaus in quantum neural network training landscapes *Nat. Commun.* **9** 4812

[12] Benedetti M, Garcia-Pintos D, Perdomo O, Leyton-Ortega V, Nam Y and Perdomo-Ortiz A 2019 A generative modeling approach for benchmarking and training shallow quantum circuits *npj Quantum Inf.* **5** 45

[13] Wang S, Fontana E, Cerezo M, Sharma K, Sone A, Cincio L and Coles P J 2021 Noise-induced barren plateaus in variational quantum algorithms *Nat. Commun.* **12** 6961
[14] Cerezo M, Sone A, Volkoff T, Cincio L and Coles P J 2021 Cost function dependent barren plateaus in shallow parametrized quantum circuits *Nat. Commun.* **12** 1791
[15] Tang E 2019 A quantum-inspired classical algorithm for recommendation systems *Proc. 51st Annual ACM SIGACT Symp. on Theory of Computing, STOC 2019* (New York: Association for Computing Machinery) pp 217–28

www.ingramcontent.com/pod-product-compliance
Lightning Source LLC
Chambersburg PA
CBHW080735150926
51501CB00005B/168
*9780750349505*